Arthur Hellyer's All-Colour Book of Indoor and Greenhouse Plants

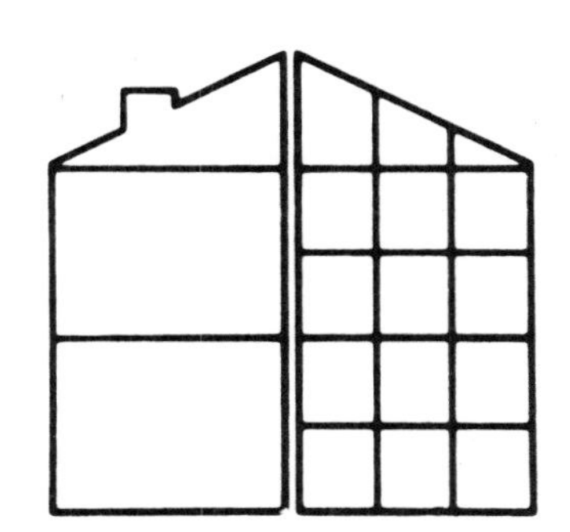

HAMLYN

London · New York · Sydney · Toronto

Acknowledgements

The Editor is grateful to Bradley-Nicholson Limited for lending the illustration of the Shilton natural gas greenhouse heater used on page 34, and would also like to thank the following for providing the remaining photographs used in this book: Bernard Alfieri, *Amateur Gardening*, Pat Brindley, Kenneth Burras, Robert Corbin, John Cowley, Ernest Crowson, W. F. Davidson, Mrs H. Hodgson, Anthony Huxley, Elsa Megson, Frank Naylor, Sheila Orme, Robert Pearson, Ray Procter, Harry Smith and D. Wilridge.

First published in 1973 by
The Hamlyn Publishing Group Limited
London · New York · Sydney · Toronto
Astronaut House, Feltham, Middlesex, England
Second impression 1977

Filmset in Great Britain by
Filmtype Services Limited, Scarborough
Printed and bound in the Canary Islands by
Litografia A. Romero, S.A. Santa Cruz de Tenerife,
Canary Islands (Spain)

ISBN 0 600 34403 7

D. L. TF. 59 - 74

Contents

Introduction

Four centuries ago gardeners invented the orangery as a means of keeping the newly introduced orange trees alive in climates too cold for them to survive out of doors in winter. By comparison with modern greenhouses orangeries were inefficient structures in which to grow plants, but they were usually handsome buildings adding materially to the appearance of many a fine estate and they performed their limited function satisfactorily. Well built with solid stone or wooden walls and tiled, slated or leaded roofs, they were, as a rule, provided with windows on one side only, that which received most sunshine. If they were heated at all, it was by flues constructed in the walls, and the amount of heat lost must have been very great, but then it was only the wealthy who could afford to contemplate growing oranges.

We have moved a long way since those days. As exotic plants more tender and exacting than the orange were introduced, gardeners had to devise better structures that would admit more light and could be heated and ventilated more efficiently. They were assisted by structural innovations such as wrought-iron glazing bars and cast-iron water pipes, and in the 19th century they constructed many large and often beautiful winter gardens, conservatories and greenhouses in which the largest and most exacting plants could be grown to maturity. These houses were warmed by water flowing to and from boilers by the natural tendency of hot water to rise and of cold water to sink. To this day this thermo-siphon principle is still used to heat many privately owned glasshouses, though for commercial purposes it has been superseded by more efficient systems, such as pump-circulated water, high pressure steam and various systems of warmed (and sometimes humidified) air blown into the houses.

Electrical heating systems have also been developed and, though usually considerably more expensive to run than those operating on solid fuels or oil, they are often relatively cheap to install and are always extremely easy to control. They have been of major importance to amateur gardeners with small to medium size greenhouses in which convenience rather than cost effectiveness is the first consideration. The owner of an electrically heated greenhouse does not have to order fuel, has no fire boxes to clean or chimneys to sweep and the whole system can be made completely automatic by fitting a simple thermostatic control.

Now electricity, so long supreme as a labour-saving source of heat, is being challenged by natural gas which, unlike the gases produced from coal and oil, gives off no harmful fumes when burned. The waste products of natural gas – water and carbon dioxide – are positively beneficial to plants and so it can be burned within the greenhouse with every unit of heat trapped where it will do most good. Gas heating is as easily controlled by thermostat as electricity, the average cost of installation is certainly no greater and running costs are considerably lower. The limiting factor, as it always has been with electricity, is the availability of supply, but at least gas has the advantage that it can be bottled under pressure and distributed to remote places.

What the future holds one can only guess, but there may well be developments in the use of nuclear heat, or in tapping the vast latent sources of heat in the earth itself. On the structural side it may well be that we shall see the natural light abandoned in favour of artificial light, on the grounds that it is cheaper to illuminate than it is to heat, and that it is cheaper to warm a building with well insulated walls than one made largely of glass or translucent plastic. Plant houses in which all windows have been eliminated are already in use and have proved their practicability, and some are also heated from natural sources thus effecting ever greater economies. Even the amateur can share in these new techniques by investing in a plant cabinet, a kind of miniature plant house designed to stand inside a room and to provide, under complete control, any type of climate that particular plants may require. It may all seem a far cry from the 17th-century orangery, yet in one way the wheel may be said to have turned full circle for the orangery was primarily an efficient insulator against outside cold and that is the major function of the 20th-century plant house and plant cabinet. The main deficiency of the orangery, its lack of light, has been rectified by developments in illumination undreamed of then.

There is also a similarity between present-day interest in garden rooms and house extensions and the old-time winter gardens and conservatories, for all have the dual purpose of being made for the use of both people and plants. We also bring plants right into our homes and, since these are usually better lighted and are almost invariably better heated than the 17th-century orangeries, we are constantly discovering that we can grow indoors a much greater variety of plants than at one time seemed probable. In most rooms it is neither lack of light nor lack of warmth that is the major limiting factor but rather lack of humidity since our requirements for comfort and those of many plants for health do not coincide. The indoor plant cabinet, its elegant predecessor the Wardian case, and that decorative oddity, the bottle

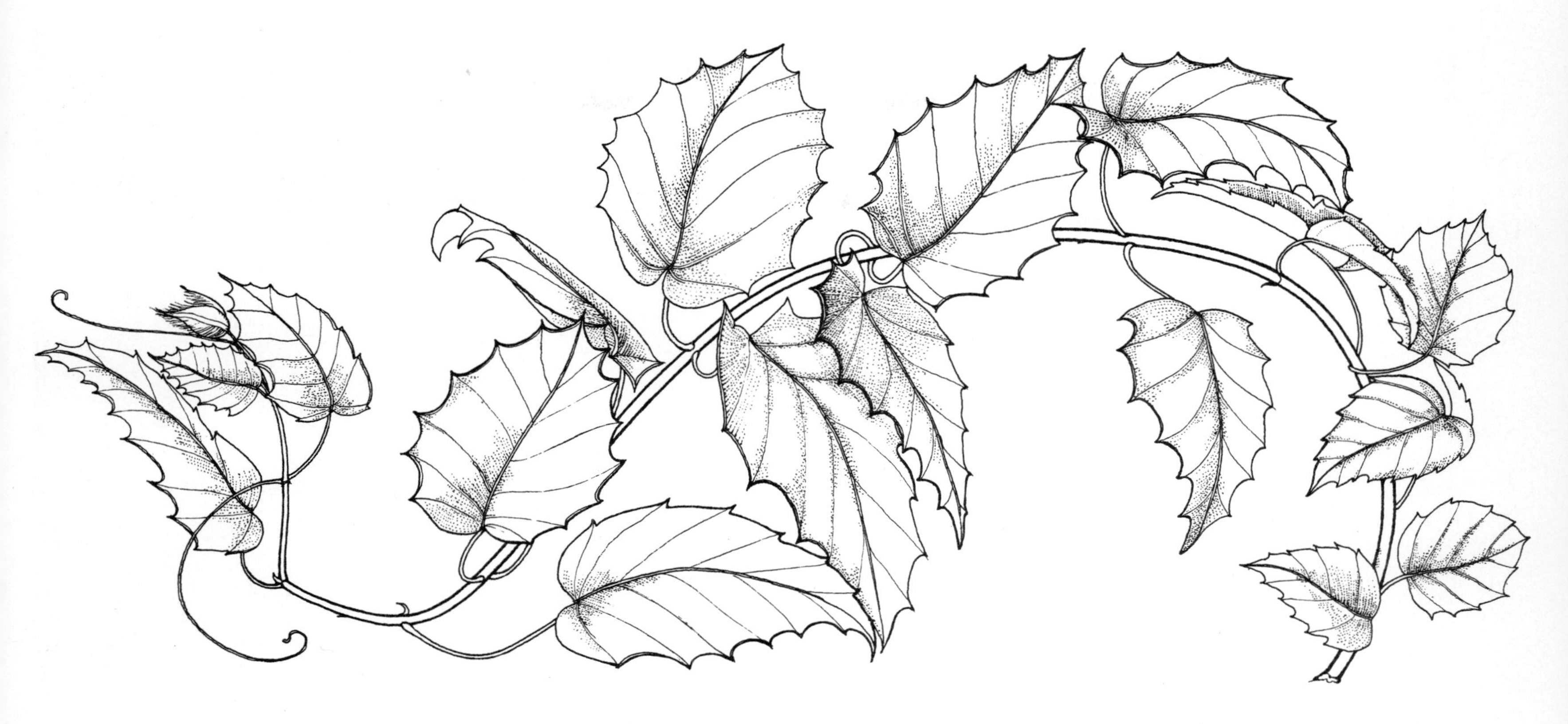

garden, are all means of overcoming this difficulty and bringing a greater variety of plants even more closely into the daily pattern of our lives.

Another invention that is having a similar effect is the humidifier. This device, essential in some manufacturing processes, is now widely used in hospitals and is creeping into offices and private houses as it becomes increasingly apparent that the very dry air induced by central heating is neither as comfortable nor as good for health as we have supposed. The day is probably not far distant when indoor humidity will be as much under finger-tip control as heating and lighting are today, and though we shall never want to live in the moisture level of a tropical forest we may well find that the ideal conditions for human beings and a vast range of temperate and sub-tropical plants are not so very different.

That the interest in house plants has come to stay is evident. It is all part of urban man's attempt to re-forge his links with nature and architects and builders are already taking it into account. The picture window, which did so much to bring house plants into favour, was developed for people rather than for plants, but the extra wide window-sill, tiled and dished or channelled to conserve moisture, is a refinement for the benefit of plants and one which could no doubt be extended in many ways. Some architects are considering the possibilities of hydroponics or other systems of soilless cultivation to extend the growth of plants in buildings, and are making provision for such installations in their designs. Lightweight materials such as vermiculite and expanded polystyrene have already proved their worth in the cultivation of some plants and could be more widely used if the supply of plant nutrients to such sterile rooting media could be made sufficiently reliable and foolproof.

Gardeners often talk as if temperature, moisture and light intensity were the only three considerations that matter when providing artificial climates in which to grow plants. In fact there is a fourth which for certain plants is just as vital. This is the relative length of day to night, for this controls the growth and flowering sequences of many plants. In a broad way plants can be divided into three groups: short-day plants which keep growing until nights exceed a certain length when they begin to produce flower buds; long-day plants which mainly grow when days are of medium length and start to flower only when they become long; and plants which have no clear photoperiodic pattern (the scientific name given to this phenomenon). The chrysanthemum is a typical short-day plant, the fuchsia a long-day plant, but temperature can complicate matters. The popular Christmas Cactus (page 25) is a short-day plant in fairly warm temperatures, but is unaffected by day length in temperatures below 10° C. (50° F.). If it is grown in a normally heated room which is artificially lighted most evenings in autumn and winter it may not flower at all, since the night is shortened (or the day lengthened) by the lighting. But if it is covered each evening and left in the dark for at least 12 hours it will flower quite normally. Similar effects of inducing growth or flowering at will can be produced by changing the colour of the light and again all kinds of exciting developments appear possible as our knowledge of plant behaviour increases.

To conclude on a less technical note, it is worth observing that the success of the orangery was dependent on the fact that the orange bushes only needed protection in winter. During the summer, when they were making their growth and needed much more light than the orangery could provide, they stood out of doors as objects of beauty as well as proof of their owner's wealth and horticultural skill. Today some house plants are placed out of doors in summer and many more would be if their owners only realised the improvement this could make to their health. Like the oranges of old, many house plants would make handsome specimens to decorate a patio or sheltered terrace from June to September. Provided they were assured such an annual vacation outdoors, many more species could be added to the already extensive list of plants to be grown indoors. Many of the popular grey-leaved plants, such as the various species of euryops, numerous very beautiful artemisias, silvery helichrysums and slightly tender olearias, fit perfectly into such an in-and-out routine since they are plants that appreciate dry air in winter but welcome all the light and warmth that is available in summer.

House Plants

As the interest in house plants increases so the range of available plants grows. In part this is due to the competition of commercial growers anxious to cash in on the potentially highly profitable market. The more enterprising of these are constantly searching for new plants that have the necessary toughness to survive the conditions of an average living room, office or shop. Sometimes these discoveries are really new in the sense that they are plants not previously cultivated, or grown on so limited a scale as to be virtually unknown. More often the discovery is simply that a plant previously believed to be unsuitable for room cultivation can be so grown provided it is given the right treatment.

But there is another reason why the list of house plants is now so much longer than it used to be and is constantly growing. This is that rooms are on average much better lighted and warmed than they used to be. Big picture windows and central heating have done more for house plants than anything else though they do bring new problems of their own, the windows actually encouraging an excessive amount of heat on sunny days and the heating aggravating the already over-dry atmosphere of living rooms. Decorative Venetian blinds can take care of the first problem and automatic humidification can look after the second. Already the use of both these devices is increasing rapidly and they may soon be taken as much for granted in the modern home as central heating is today. A few architects make positive provision for house plants when preparing plans for new buildings and this may well prove popular and be more widely practised.

The qualifications of a first-rate house plant are that it must be able to grow and look happy in a relatively dry atmosphere and poor light, and that it must be suitable for cultivation in pots of reasonable size. For special places one can go outside these limits but they remain essential for the really big sellers.

Plants that fulfill these conditions may be further grouped under three headings: one, the very easy plants that can be grown in rooms in which the temperature sometimes falls as low as 7° C. (45° F.); a second, those requiring a rather higher temperature, rarely falling below 13° C. (55° F.), and the third, tropical plants which enjoy temperatures of 18° C. (65° F.) or more all the year. It will be observed that even the last come well within the range of what would be regarded as average temperatures in any centrally heated room. Most house plants will appreciate being grown in the lightest part of the room, though not necessarily in a window where they will get a great deal of direct sunshine. A few will survive even far removed from windows in poorly lighted corridors. All dislike being placed near fires or stoves which are only used intermittently or in a position exposed to draughts.

Care of House Plants

The leaves of evergreen pot plants should be sponged occasionally, using either one of the proprietary preparations or a home-made mixture of milk and water

House plants may be grown in John Innes or a peat-based potting compost. Commercial growers are tending increasingly to use peat composts and it is in these that plants are likely to be growing when purchased. When potting on into a larger pot as the plant increases in size it is wise to use compost of a similar type to that in which it is already growing.

Water house plants whenever the soil shows signs of drying out, which may be daily in warm rooms when the plants are growing, and feed once every two or three weeks from May to August with a liquid plant food. Large evergreen leaves should be sponged with water two or three times a week in spring and summer, with a little white oil emulsion added if red spider mites are troublesome.

If it can be managed, pack damp sphagnum moss around the pots or plunge them in containers of moist peat, both to keep the air around the plants moister and to slow the drying of the soil. Moss or peat can be watered daily to keep it moist.

Only re-pot when pots become overfilled with roots and then into a pot which is only slightly larger. This should be done in spring and some of the old compost removed and replaced with fresh.

Plant Cases

Plant cases provide a greater control of growing conditions and they can, therefore, be used to grow a much wider range of plants than would be possible in the atmosphere of a room

The Wardian case was popular in the 19th century as a means of growing ferns and some other moisture-loving plants indoors. It is really a miniature greenhouse in which a stable atmosphere can be maintained and is small enough to be stood on a table.

Wardian cases are still occasionally seen as period pieces, but their place has largely been taken by more elaborate cabinets of various sizes and designs, usually equipped with their own electric heating and lighting. Because temperatures, humidity and illumination can all be controlled, a much wider range of plants can be grown than would be possible in the atmosphere of the room itself, and really the main limitation is imposed by the size of the case and what it can conveniently contain. Such cabinets are excellent for cultivating small flowering plants such as saintpaulias and impatiens and may also be used for some small-growing kinds of orchids.

Plants growing in plant cabinets should be treated in much the same way as plants in greenhouses.

Do not use artificial lighting in winter to extend day length to more than 12 hours unless it is known that the plants grown will respond well to this. The growth and flowering periods of some plants are controlled by the ratio of day to night length and variations can be upsetting.

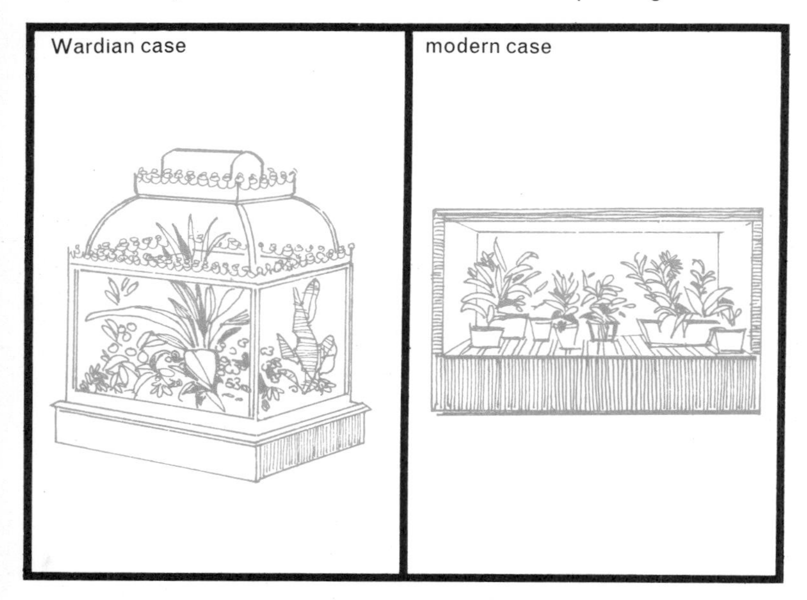

Bottle Gardens

If care and patience are exercised in creating bottle gardens, they will make very attractive features which will require little attention for several years

Bottle gardens are usually created in large carboys or specially made bottles of a similar globular shape. Place two or three inches of moist peat-based compost in the bottom of the bottle, running it through a paper funnel to avoid soiling the sides, and insert small house plants through the neck, planting them with the aid of long sticks or specially made wooden tongs. It is best not to stopper the mouth of the bottle as a slight circulation of air is desirable to prevent condensation forming, though the rate of drying out will be very slow. Under normal room conditions watering will only be necessary three or four times a year. However, the soil must not be allowed to dry out and a moderate quantity of water should be applied as soon as it shows signs of doing so.

Provided suitable slow-growing plants are chosen and they are not fed in any way, a bottle garden can remain in good condition for several years. Eventually, however, plants will become overcrowded or some will fall into ill health and then the bottle will have to be emptied and the process started again.

Some of the smaller ferns grow well in bottle gardens, suitable kinds being *Adiantum cuneatum*, a maidenhair fern; *Pellaea rotundifolia*, and silver-variegated varieties of *Pteris* such as *argyraea* and *victoriae*. Other good plants are the variegated acorus, cryptanthus, fittonia, maranta, peperomia and sonerila.

Bulbs in Bowls

Mixed hyacinths growing in a pottery bowl are a bright addition to the home in the winter months, and will quickly fill a room with their 'heady' scent. Specially prepared bulbs will flower at Christmas

It is convenient to grow bulbs for the home in bowls without drainage holes, so that no water can trickle out to damage furniture, fabrics, etc. Ordinary soil and potting composts quickly become sour in such bowls and special bulb fibre must be used. This can be purchased ready for use or prepared at home with 6 parts by bulk of medium grade horticultural peat, 2 parts crushed shell and 1 part crushed charcoal, well mixed.

Thoroughly moisten the bulb fibre before use. Place a layer of fibre in the bottom of the bowl, set bulbs shoulder to shoulder on this and barely cover with more fibre. Water well, and place in a cool, dark place (such as a cupboard, but not an airing cupboard which will be too hot) for from eight to ten weeks so that roots may be formed freely. Then bring bowls into a light, moderately-warm place. If in a window, give each bowl a quarter turn daily so that the growing plants receive equal illumination all round. Support flower stems with split canes or thin stakes placed around them with encircling ties of green twist.

After flowering, plant bulbs carefully outdoors where they can complete their growth. Do not use the same bulbs in bowls for two years running.

All hyacinths, including, 'prepared' ones, and some daffodils such as Cragford and Geranium succeed especially well in bowls.

Popular Flowering Plants

Impatiens or Busy Lizzie is a popular indoor flowering plant which will do well when grown in a light sunny window, though it should not be allowed to dry out as the buds will drop

Some flowering plants, dealt with more fully elsewhere in this book, make good pot plants for well-lighted rooms. These include cyclamen, impatiens (Busy Lizzie), greenhouse azaleas, pelargoniums, hibiscus and anthurium.

Of these, impatiens and anthurium are best fitted to remain indoors permanently. Impatiens grows quite well in shade, is not too seriously affected by the rather dry air of most rooms, and is capable of flowering all the year.

Cyclamen need a marked rest in summer and are then much better stood outdoors, preferably shaded from direct sunshine. They need not be brought inside again until late September.

Azaleas are also much better out of doors from May to October. They dislike dry air while making their growth and flower buds and should be well watered, frequently syringed with water and occasionally fed with weak liquid fertiliser while outside.

Pelargoniums of all kinds need more light, especially in summer, than is normally available in rooms and may be stood out of doors from June to September in a sunny place. Hibiscus can be treated in the same way.

The best anthurium for rooms is *A. scherzerianum*, the Flamingo Flower, as it puts up with dry air better than the larger-flowered *A. andreanum*. It should be grown in a light window but not in full sun and should never be put outdoors.

Hibiscus rosa-sinensis

Aglaonema and Others

The attractive flowers of *Spathiphyllum wallisii* have given it the popular name of White Sails. It should be protected from direct sunlight and should not be allowed to dry out

Aglaonema and spathiphyllum belong to the arum family and may produce arum-like flowers which add to their attractiveness. *Aglaonema treubii*, the kind usually grown as a house plant, has spear-shaped leaves, dark green splashed and banded with lighter green and the flowers are white. It likes a warm, rather moist atmosphere, a minimum temperature of 15° C. (60° F.), a fair amount of water in spring and summer but the soil only just moist in autumn and winter.

Spathiphyllum wallisii also has white flowers which are produced much more regularly and freely so that the popular name of the plant is White Sails. The leaves are similar in shape to those of the aglaonema but dark shining green without other colouring. It will succeed in temperatures 2 to 5° C. (5 to 10° F.) lower than *Aglaonema treubii*, likes shade, requires regular feeding from May to August and is best divided and re-potted each spring as it grows rapidly and soon exhausts the soil. Re-pot in John Innes or peat-based potting compost.

Related to these, though totally different in appearance, is *Acorus gramineus variegatus*. This has narrow rush-like leaves banded with white. It is hardy and can be grown out of doors in damp soil, but is also a splendid little plant for a bottle garden, preferably planted in peat-based potting compost.

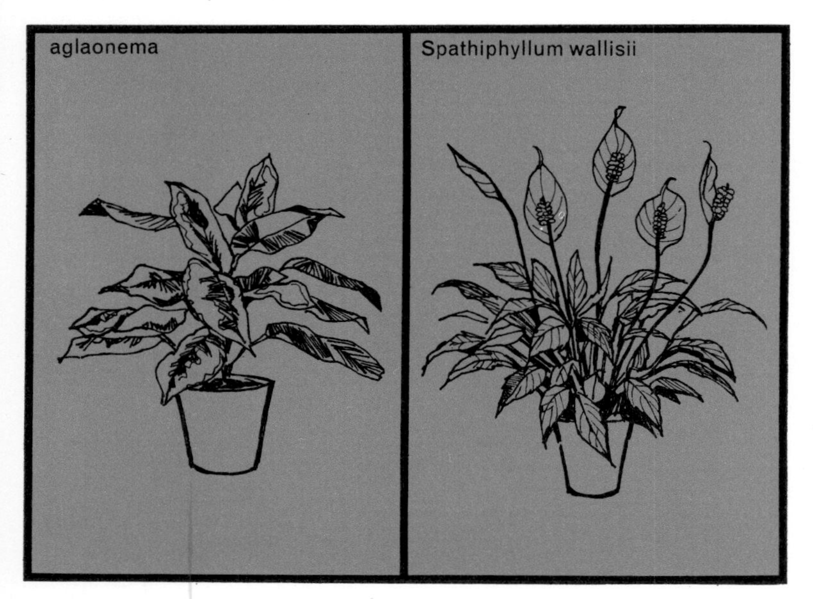

Aphelandra and Others

Codiaeums (top left) will make handsome house plants for short spells provided these are followed by periods of recuperation in a warm greenhouse. The leaves vary greatly in shape and colour

Aphelandra squarrosa louisae is known as the Zebra Plant because of the pattern of striping on its leaves, white transverse veins on a dark, shiny green leaf. The spikes of closely packed yellow flowers have a curiously wooden look and are handsome but the plant is chiefly valued for its leaves. It likes a fairly light position and a minimum winter temperature of 13° C. (55° F.).

Pachystachys lutea has similar flower spikes but its leaves are plain green. It is treated in the same way, although it is really better grown in an intermediate or warm greenhouse.

Sanchezia nobilis is also related and has varieties with yellow- or white-veined leaves. It requires similar treatment to aphelandra.

Crotons, also known as codiaeums, are also grown for their coloured leaves which may be broad or narrow, smooth or twisted in a great variety of colours, including green, yellow, rose, red, copper and bronze. They are amongst the most beautiful of foliage plants and grow well in warm, well-lighted greenhouses, but are not so easy to manage in rooms, though they make excellent house plants for short periods, interspersed with periods of recuperation in a greenhouse. They like light, fairly warm conditions and dislike widely fluctuating temperatures and draughts. They should be fed regularly from May to August and be well watered.

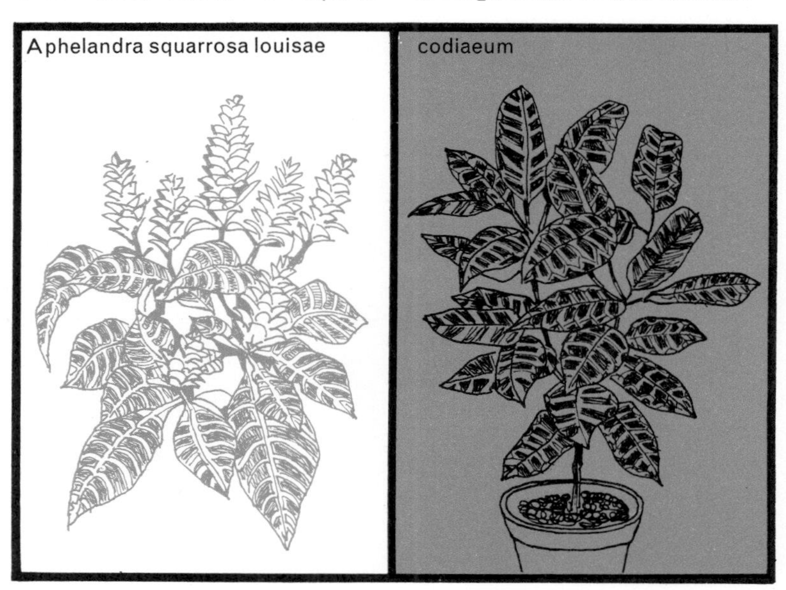

Aralia and Others

House plants flourish in a moist atmosphere, and many plants will benefit if sprayed lightly with water from time to time. Here, *Fatsia japonica* is being given this treatment

Aralia elegantissima is sometimes called the Spider Plant because of the spidery appearance of its very narrow, toothed, almost black leaves arranged in a fan like those of a Horse Chestnut. The young leaves are coppery red. It needs a light place and a minimum winter temperature of 13° C. (55° F.). Red spider mites are partial to it and since sponging of these narrow-toothed leaves is difficult they should be sprayed occasionally with derris or white oil emulsion. The plant is also known as *Dizygotheca elegantissima*.

Fatsia japonica is known as the False Castor Oil Plant, and is occasionally called *Aralia sieboldii*. It has large, shining, laurel-green leaves cut into broad lobes and it can be grown in quite unheated rooms – even out of doors – and in very poor light. There is a variety with leaves bordered with cream, but it is not quite so easy to manage.

Fatshedera lizei is a hybrid between fatsia and an ivy (hedera). It has leaves very like those of *Fatsia japonica* but smaller, and the plant makes long, flexible stems which need to be trained to a stake or trellis. There is a variety with cream-edged leaves. Treatment is the same as for fatsia.

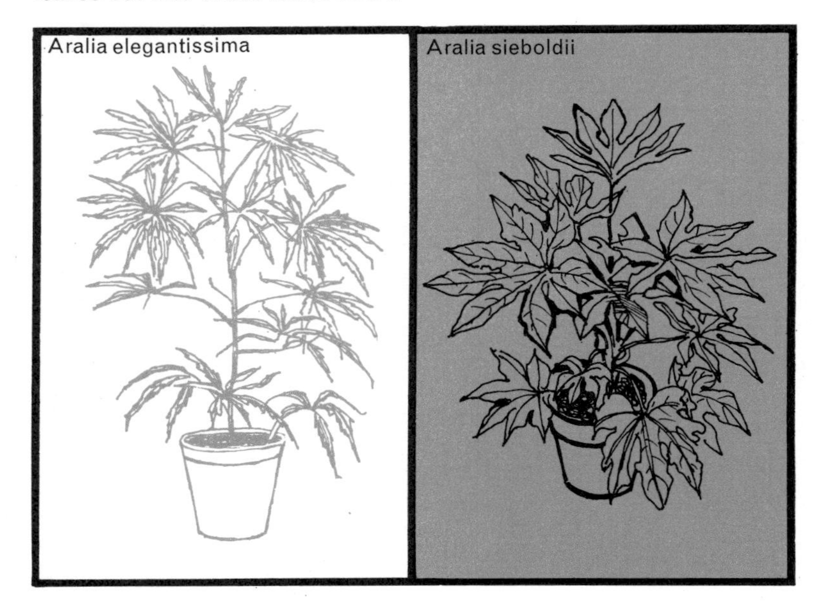

Araucaria and Others

Araucaria excelsa, the Norfolk Island Pine, is a delightful and easy-to-grow house plant. If near a window it should be turned occasionally to prevent uneven growth

Araucaria and aspidistra are old-fashioned and easily grown house plants. *Araucaria excelsa* is known as the Norfolk Island Pine, but is more like a stiffly branched fir tree. It has dark green, needle-like leaves on horizontal branches arranged in regular whorls to form a pyramid and in this it resembles the Monkey Puzzle Tree, *Araucaria imbricata*, to which it is related. It likes a light position and does not mind a minimum temperature of 7° C. (45° F.). If grown near a window it should be turned periodically to prevent uneven growth.

The aspidistra has broadly lance-shaped deep green leaves, which are striped with white in the variegated variety. It will grow in quite poorly lighted rooms and will withstand a lot of mismanagement. It can be grown in soil or peat composts and should be watered fairly freely in spring and summer, moderately in autumn and winter. The shine of the leaves is improved by regular sponging with water containing a few drops of milk. Overgrown plants can be split into several pieces in spring, each of which can be re-potted and will soon make a good plant.

***Ardisia crispa*, described more fully on page 88, is sometimes grown as a house plant and is useful for its shiny evergreen leaves, even if, under room conditions, it fails to produce good crops of scarlet berries. It should be given all the light possible.**

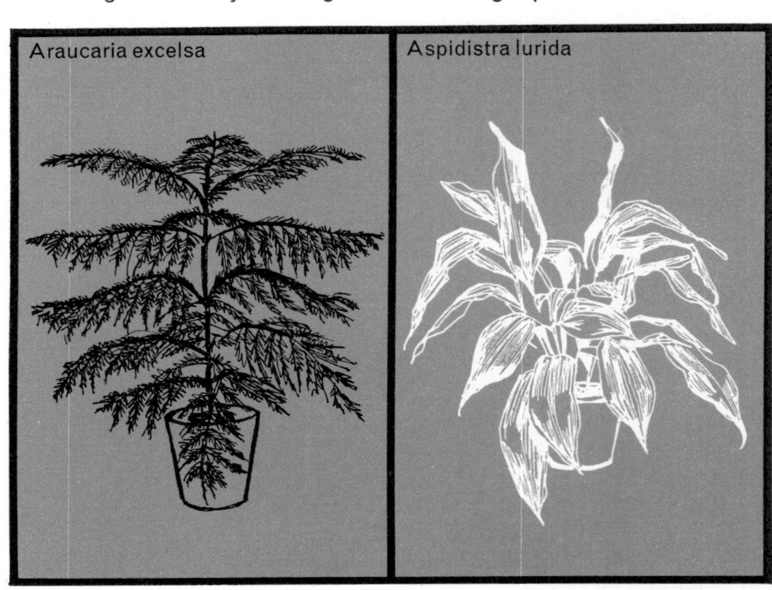

Beloperone and Others

Hoya carnosa is very reluctant to flower when grown indoors but in its variegated form it makes an attractive foliage plant. It needs to be trained against a stake or trellis

Beloperone guttata is known as the Shrimp Plant because of the vaguely shrimp-like appearance of its curling russet-pink flower heads. It is one of the few flowering plants that will do well indoors and it flowers for a long period in summer and autumn. It will thrive in any room with a minimum temperature of 7° C. (45° F.), but is best placed in or near a window as it likes light. Keep moderately watered at all times, more freely in summer, and feed from May to August.

Hoya carnosa is known as the Wax Plant because the little clusters of fragrant white flowers look as if they are made of wax. However, the flowers are not freely produced indoors and this plant is usually grown in one of its variegated forms for the fleshy dark green leaves margined with cream or with a yellow centre.

Hoya carnosa will grow in a room with a minimum temperature of 18° C. (65° F.). It is a climber which should be trained around a stake or over a little trellis and it needs a fairly light position.

Grevillea robusta, described more fully on page 61, is often grown as a house plant. It tends to grow too fast and may have to be discarded after a while, but young plants are readily raised from seed. It should be given a reasonably light place.

Bromeliads

Despite its exotic appearance neoregelia is an easy and thoroughly reliable house plant to grow. The central cup or 'vase' should be kept filled with water

The pineapple is a bromeliad and it is occasionally grown for ornament in well-heated greenhouses, but there are other more decorative members of the family. All have stiff, strap-shaped or sword-like leaves arranged in rosettes often around a central cup, referred to as a 'vase', which collects water and helps to sustain the plant. Among the best are: aechmea with mottled leaves and stiff pink and lavender or scarlet flower heads; cryptanthus with flat, starfish-like rosettes of cream, green or golden-brown leaves; neoregelia and nidularium with green and cream leaves turning to pink or scarlet around the vase and vriesia with banded leaves and quill-like scarlet flower spikes.

Grow bromeliads in the smallest pots that will contain their roots. Pot them in a mixture of equal parts sphagnum peat, sharp sand and osmunda fibre. Grow in a temperature of 18° C. (65° F.) or more, which may drop occasionally in winter to 13° C. (55° F.) but no lower, and keep away from direct sunshine. Maintain as moist an atmosphere as possible. Water freely in spring and summer, moderately in autumn and winter, always keeping the central cup or vase filled with water, and feed with very weak liquid manure about once a fortnight in summer. Increase by removing offsets in May.

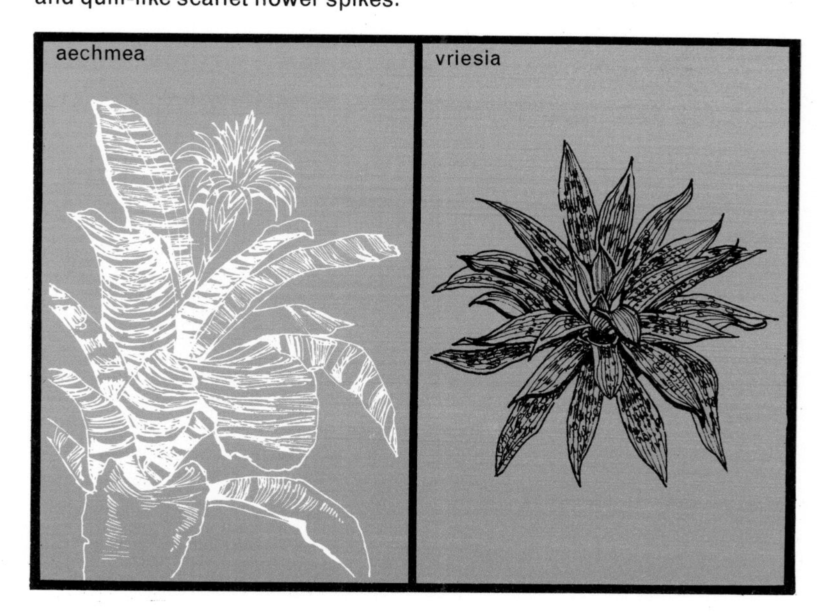

Calathea, Maranta

Warm, humid conditions are ideal for marantas. To maintain the humidity around the leaves it is often necessary to pack damp moss or moist peat round the pots. Keep the plants out of direct sunshine

These are related plants, grown for their attractively marked foliage. All are fairly small, compact plants thriving in warm, rather humid conditions, for which reason they are excellent plants for bottle gardens. However, they can also be grown in rooms in which the temperature does not fall below 13° C. (55° F.). It is desirable to pack damp moss around the pots or to plunge them in moist peat to maintain humidity around their leaves. Keep them out of direct sunshine which may scorch the leaves and water moderately at most times, but a little more freely in summer.

The leaves of calatheas are oval or oblong in various shades of green on top, regularly blotched with darker green, and shades of purple beneath. *Calathea makoyana,* also known as *Maranta makoyana,* is one of the most beautiful and is often known as the Peacock Plant.

Maranta leuconeura is a shorter plant, very variable in the colour of its leaves which may be light green with dark green or red blotches and with white or crimson veins.

All calatheas and marantas thrive in peat-based potting composts.

Calathea makoyana

Chlorophytum and Others

Tradescantia is an excellent plant to grow round the edge of a container from which it will hang attractively. It grows almost anywhere and does not require much attention.

Chlorophytum comosum variegatum

Rhoeo discolor

Zebrina pendula

These are among the easiest of house plants to grow. They will thrive with little light and quite low temperatures, certainly down to 7° C. (45° F.) or even a little less.

Chlorophytum comosum variegatum makes a large tuft of long narrow leaves banded with creamy white and the plant produces long, arching stems bearing plantlets which can be used for propagation. The leaves and stems of *Rhoeo discolor* are purple and the habit of the plant sprawling.

Setcreasea purpurea is allied and is known as Purple Heart. It has purple leaves and stems, grows upright, and can be cut hard back if it becomes untidy.

Tradescantia fluminensis and *Zebrina pendula* are so much alike that they share the same popular name, Wandering Jew. They are trailing plants with oval, light green leaves striped with silver, cream and pink. In zebrina the leaves are purple underneath. They are excellent plants to grow round the edge of a container, over which they will hang, but they can be grown almost anywhere in John Innes or peat-based potting compost, and it is almost impossible to kill them even by over or under-watering.

Water freely in spring and summer, moderately in autumn and winter. Divide the plants when they get too large or take cuttings of firm young stems. Raise young plants of chlorophytum by pegging down the plantlets to the soil.

Cissus and Others

The rhoicissus (right) is a fine house plant which will tolerate all but the worst conditions. The leaves are made up of three leaflets which are bronzy when young becoming deep green as they age

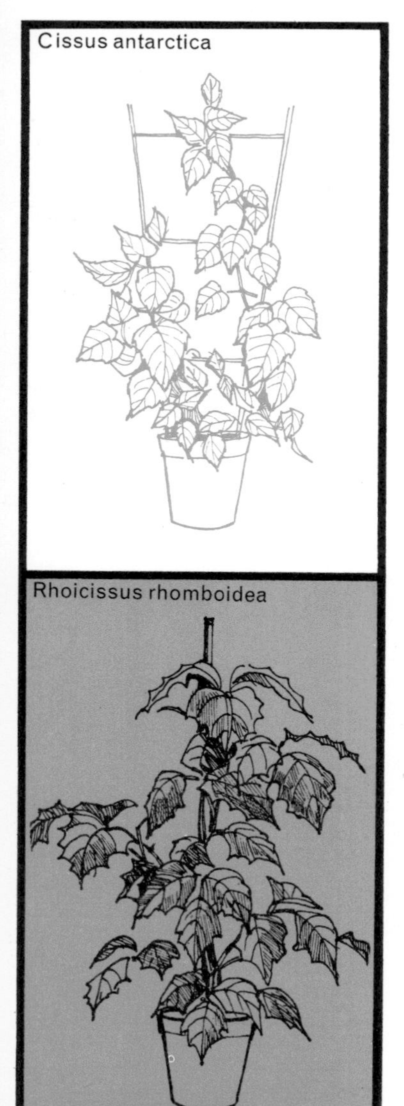

These are climbing plants with something of the character of vines–indeed, the popular name of *Cissus antarctica,* the Australian kind commonly grown as a house plant, is Kangaroo Vine and of *Rhoicissus rhomboidea*, Natal Vine. Both support themselves by means of tendrils and should be given stakes or a little trellis up which to climb.

The leaves of *Cissus antarctica* are heart shaped, often toothed and lustrous green, and the stems are reddish purple. Each leaf of rhoicissus is made up of three leaflets, bronzy at first but turning deep shining green as they age.

Allied plants are *Cissus striata* and *C. sicyoides*, both of which have leaves composed of several separate leaflets, quite small and densely produced in the first named, and *Tetrastigma voinierianum*, also with compound leaves but composed of very large leaflets. It grows very vigorously and is only suitable for growing where there is ample space.

All these plants can be grown in rooms with a minimum temperature of 7° C. (45° F.). They will tolerate poor light, but can also do well grown near or around windows. They should be watered fairly freely in spring and summer, only moderately in autumn and winter. Tips of growing shoots can be pinched out occasionally in summer if a more bushy habit is wanted.

Cordyline and Others

The narrow sword-like leaves of cordyline (right) may be coloured green to deep purple and are sometimes handsomely variegated with cream or white. They thrive in a minimum temperature of 10° C. (50° F.)

These are all plants which may be referred to popularly as 'palms', though they are not really palms nor related to the palm family. But their large rosettes of long, narrow, strap-shaped leaves give them an exotic and slightly palm-like appearance. They make distinctive and decorative pot plants for the house.

There are numerous varieties of cordyline with leaves differing in colour from green to deep purple, sometimes variegated with cream or white, but all will thrive in rooms with a minimum temperature of 10° C. (50° F.).

Dracaenas are also numerous and differ not only in the colour but also in the width of their leaves. All are known by the popular name of Dragon Plant but some require more warmth than others. *Dracaena deremensis warneckii* and *D. fragrans massangeana* enjoy a minimum temperature of 15° C. (60° F.), whereas *D. godseffiana* and *D. sanderiana* will be quite content with a minimum of 13° C. (55° F.).

Pandanus veitchii is popularly called the Screw Pine (after the pineapple, not the pine tree). The leaves are fairly broad, green with a white margin, and it prefers fairly warm conditions and a minimum temperature of 15° C. (60° F.).

Pleomele reflexa has yellow leaves striped with green.

All these plants like fairly light places, should be watered freely in spring and summer, but very moderately in autumn and winter.

Dieffenbachia, Syngonium

Dieffenbachia demands a minimum temperature of 15° C. (60° F.) and prefers a shady position in the room. The leaves should be sprayed with water frequently in the summer

Dieffenbachia picta, the kind commonly grown as a house plant, has large leaves carried around a central stem in a somewhat tree-like fashion. The leaves are variously blotched with cream or yellow according to the variety; indeed, in some there is more variegation than basic green colour. All like shade and need warm rooms with a minimum winter temperature of 15° C. (60° F.). They should be watered rather freely in spring and summer, just sufficiently to keep the soil moist in autumn and winter, and should be fed from May to August. Leaves should be sprayed frequently in summer.

Syngoniums, like dieffenbachias, belong to the arum family, but are climbing plants. The leaves are shaped like an arrowhead and are either dark, shining green or variegated with lighter green along the veins. All kinds require the same conditions as dieffenbachias but must be provided with some support to which their long stems can be tied, the best choice being moss-covered wire or cork that can be kept moist. The popular name for this plant is Goosefoot Plant and syngonium is often sold as nephthytis.

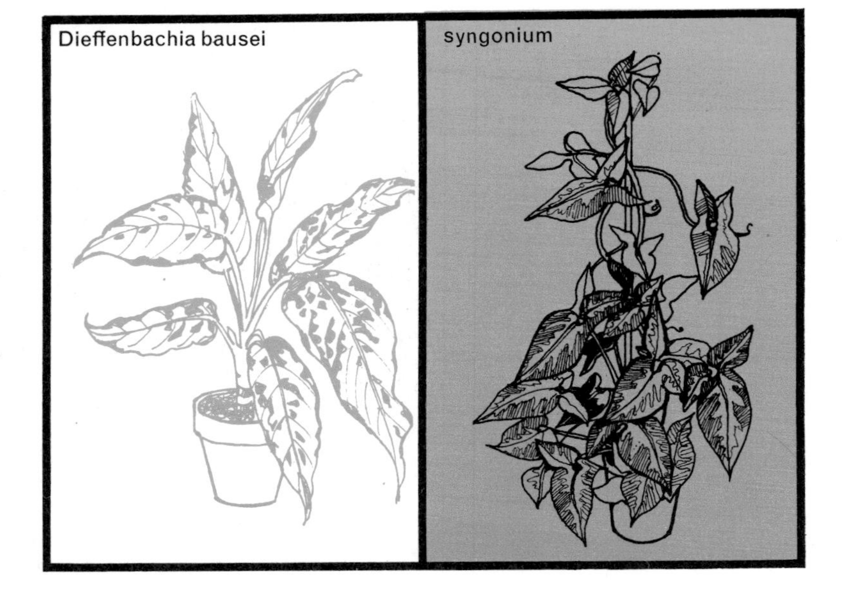

Ferns

Asplenium bulbiferum has finely divided fronds which carry tiny plantlets on the upper surface. This is a good fern for the home provided the air does not get too dry

By no means all ferns make good room plants as the atmosphere is too dry for them, but some kinds will survive for long periods, especially if damp moss is packed around the pots to keep the air moist.

Among the most satisfactory kinds are the Bird's Nest Fern, *Asplenium nidus,* with broad, shining dark green fronds; the Holly Fern, *Cyrtomium falcatum,* with rather large holly-green segments to its fronds; and the Ribbon Fern, *Pteris cretica,* with long, narrow, ribbon-like segments which in some varieties are crested, divided or waved. Others that may be tried provided the air is not too dry are the Mother Spleenwort, *Asplenium bulbiferum,* with finely divided fronds carrying tiny plantlets on the upper surface, and the Ladder Fern, *Nephrolepis exaltata,* with ladder-like fronds and a great many varieties.

The Stag's Horn Fern, *Platycerium bifurcatum,* is peculiar in that it obtains its food from water and the air, not from the soil. For room cultivation it is grown in peat in a pot so that it can be kept moist, but the pot must be placed on its side except when being watered. Plants can be suspended on walls where their large greyish green, antler-like fronds look very handsome.

Ficus

Ficus elastica decora, an improved form of the India-rubber Plant, is a very popular house plant. Care should be taken not to overwater which can cause the loss of the lower leaves

Ficus elastica, the India-rubber Plant, is one of the most hard-wearing plants and one of the few which will grow in quite poorly lit places. It has large, dark green, shining, leathery leaves; but there are also variegated varieties, one mottled with cream, another with cream and light green. All are best grown in rooms with a minimum temperature of 13° C. (55° F.), though the green-leaved variety will stand slightly lower temperatures. All should be watered fairly freely from April to September, sparingly from October to March and should be fed regularly from May to August.

Ficus pendula and *F. radicans* are creeping plants which will climb like ivy with the aid of clinging aerial roots. Both have small rounded leaves and there is a pretty variegated variety of *F. radicans* with cream-edged leaves. Of the two, *F. pendula* is the hardier, thriving in a minimum temperature of 7° C. (45° F.) whereas *F. radicans* prefers 10 to 13° C. (50 to 55° F.). Both like shade and should be watered throughout the year sufficiently to keep the soil moist.

The variegated variety of *F. radicans* is particularly sensitive to dry air. It can be grown in a bottle garden or plant cabinet or be allowed to climb up a pillar made of wire netting packed with sphagnum moss and kept moist.

The giant of the family, *F. lyrata*, is known as the Banjo or Fiddle-leaved Fig. It is grown like *F. elastica* but needs plenty of room.

Fittonia and Others

Fittonia argyroneura has handsomely marked leaves. In common with other fittonias it enjoys warmth and humidity and is an excellent plant for the bottle garden or plant cabinet

Fittonias are low-growing perennials with ornamentally veined leaves, white on green in *Fittonia argyroneura*, pink on purple in *F. verschaffeltii*. Both enjoy warmth and moisture and are more suitable for growing in bottle gardens, Wardian cases or plant cabinets than in a drier room atmosphere. Both can be grown in J.I.P. No. 2 or peat-based potting compost, should be watered normally and kept out of strong sunshine. They can be increased by division in spring, the best season for repotting, but this is seldom required. Both can also be grown in a shady part of an intermediate or warm greenhouse.

Sonerila margaritacea is another tender perennial grown for its intricately coloured leaves, which in some forms are so heavily flecked with silver as to appear more silver than green. Underneath they are purple. It requires similar conditions to the fittonias. Increase is by cuttings of young shoots in a warm propagator or from seed sown at 20 to 24° C. (70 to 75° F.).

Hypoestes sanguinolenta is called the Polka Dot Plant because its leaves are covered in pink spots. It is easily grown from seed sown in a temperature of 15 to 18° C. (60 to 65° F.) in spring in J.I.S. or peat-based seed compost. The seedlings should be potted singly in J.I.P. No. 1 or peat-based potting compost and grown on in any reasonably light place with normal watering.

Gynura and Others

The velvety purple leaves of *Gynura sarmentosa* make it a most distinctive trailing plant. However, the ragged orange flowers are most unattractive and should be removed at bud stage

Gynura sarmentosa is a trailing perennial with soft violet-coloured stems and young leaves, the colour persisting on the underside of the old leaves. It will grow in any reasonably light place in J.I.P. No. 1 or peat-based potting compost, but should not be overwatered, particularly in winter. It can be increased by cuttings in spring or summer in a warm propagator.

Hypocyrta glabra is known both as the Clog Plant and the Goldfish Plant because of the shape and colour of its little orange flowers produced in summer. The rounded leaves are closely set on the stems, rather thick, glossy and dark green. Altogether, this is a distinctive little plant and one that can be grown in any reasonably light place, even in a window with direct sunshine. It should be grown in J.I.P. No. 1 or peat-based potting compost with normal watering and can be trimmed or pinched in spring or late summer if it gets too large. Increase is by cuttings of firm young shoots in summer in a propagator. It does not need much warmth and can also be grown in a cool or intermediate greenhouse.

Plectranthus oertendahlii is a perennial grown for its leaves, veined white on a bronze and green ground colour, and its sprays of small purplish-white flowers in summer. It requires similar treatment to hypocyrta except that it is best kept out of direct sunshine. Increase by division when re-potting in spring.

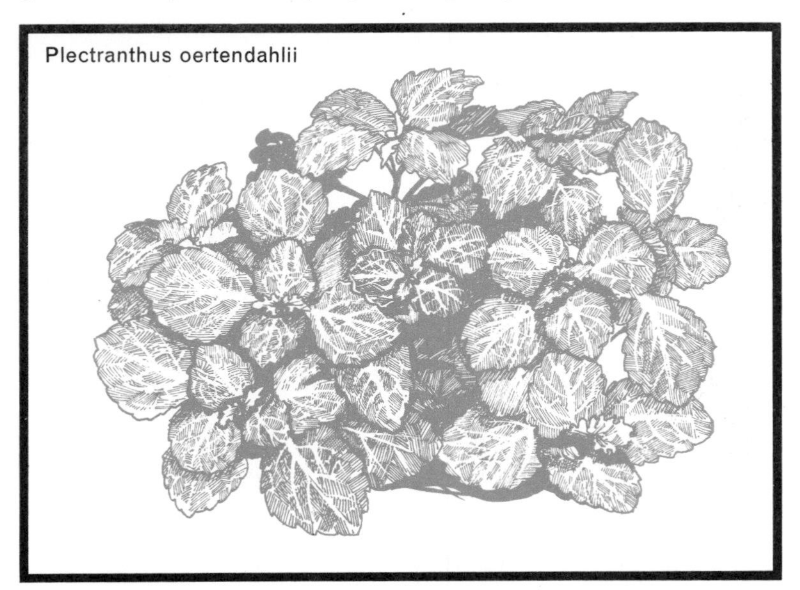
Plectranthus oertendahlii

Hedera

Ivies are very accommodating house plants for they are easy to grow and can be put to many different decorative uses. They are readily increased from cuttings

Ivy (or hedera) is one of the most easily grown plants for indoors, and can be used in many decorative ways. Ivies can be collected in the country, potted in John Innes or peat-based compost and used in the home but more decorative varieties are available in the shops. Some have small leaves which may be variegated with grey, silver, cream or yellow. Others have larger leaves or leaves of unusual shape or which change shape as they age. Yet others branch and make little bushes which need no support.

All ivies can be grown in unheated rooms. They will grow in sun or shade in John Innes or peat-based potting compost. They should be watered throughout the year, freely in spring and summer, moderately in autumn and winter. Most can be readily increased by cuttings in summer.

The Cape Ivy is a quite different plant belonging to the groundsel family. Its name is *Senecio macroglossus;* it has ivy-like leaves with a cream edge in the most popular variety, and in a light place it may produce yellow, daisy-like flowers. It requires a minimum temperature of 7° C. (45° F.), otherwise treat as for ivy. *S. mikanioides* is very similar and requires the same treatment. There is a good cream-variegated variety of *S. macroglossus*.

Monstera and Others

The distinctively cut leaves of the monstera make it the dominant plant in this arrangement. The other plants are *Begonia rex*, impatiens, dracaena and *Scindapsus aureus*

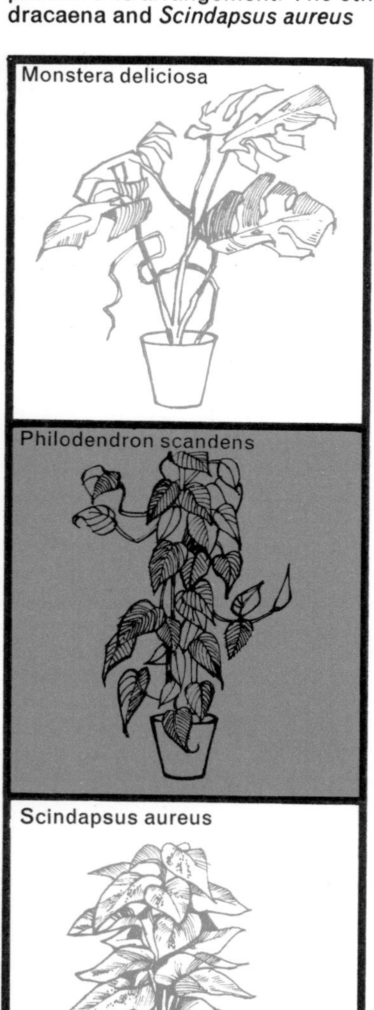

These are all related and easily grown plants, mostly climbing, which will thrive in quite poor light. Monstera is the most striking in appearance with big leaves cut and divided in a curious way which makes them appear perforated with large holes. As they grow they produce hanging aerial roots which increase the appearance of jungle-like profusion.

There are several different kinds of philodendron of which *Philodendron scandens* is the most popular. It has green heart-shaped leaves and is a climber. So is *P. hastatum* with broadly arrow-shaped leaves, but *P. bipinnatifidum*, with very large, doubly divided leaves is a bushy, non-climbing plant.

Scindapsus aureus is much like *Philodendron scandens* but its leaves are blotched with yellow.

All these plants can be grown in rooms with a minimum temperature of 13° C. (55° F.) and monstera, *Philodendron scandens* and *Scindapsus aureus* in temperatures 3 or 6° C. (5 or 10° F.) lower. All should be watered fairly freely in spring and summer, moderately in autumn and winter and should be fed from May to August.

Palms

Howea belmoreana is one of the popular palms, and will live for many years provided it is given plenty of water during the summer months and is potted on regularly

There are many different kinds of palm, most of which can be grown as pot plants, but many are so large that they are only suitable for large rooms and halls. Most succeed better in a shady greenhouse than in a room since they enjoy a moist atmosphere, but all can be used for room decoration for periods of a few weeks or even months.

The best kind for permanent cultivation indoors is *Neanthe bella*, the Dwarf Palm. This is so slow growing that it can be planted in a bottle garden, and it will grow in sun or shade. Water freely in spring and summer, sparingly in autumn or winter. Sponge the leaves occasionally to keep them glossy and if they are attacked by scale insects (like minute limpets) sponge or spray the leaves with white oil emulsion.

Phoenix roebelinii, a dwarf relation of the Date Palm, also makes a useful pot plant most suitable for light rooms or greenhouses.

Others are *Howea* (or *Kentia*) *belmoreana* and *Trachycarpus fortunei*, the last sufficiently hardy to be planted outdoors in areas where frosts are not severe or prolonged when it becomes too large for room or greenhouse cultivation.

Cocos weddeliana is a small palm with very graceful leaves. It is so slow growing that it is often planted in bottle gardens and since it enjoys a warm damp atmosphere is really better there than in the open room.

Peperomia, Pilea

These two peperomias, *P. rotundifolia* and *P. tithymaloides* (in foreground), are recent introductions as house plants, the former producing unusual flowers

There are several different kinds of peperomia, all small plants with rather fleshy, ornamental leaves. *Peperomia caperata* has wrinkled, velvety looking green leaves marked with purple and grey. *P. hederaefolia* has grey-green leaves with white veins; *P. sandersii*, also known as *P. argyreia*, has silvery leaves veined with dark green; *P. magnoliaefolia* is cream or pale yellow with a central splash of grey-green which widens as the leaves age.

All need rather careful watering as they are inclined to rot if kept too wet, but they like a moist atmosphere and shade. They enjoy a minimum temperature of 13° C. (55° F.) and seldom need re-potting as they make little root, getting much of their food from the air. All are good plants for bottle gardens.

Pileas are also small but less decorative. *Pilea cadierei nana* is the best, with dark green leaves streaked with silvery grey, for which reason it is known as the Aluminium Plant. *P. microphylla* is also known as *P. muscosa* and its popular name is Artillery Plant because if it is disturbed when in flower it emits puffs of pollen. The leaves are small and crowded making the plant look like a fern or moss.

Both these plants are very easily grown in any room with a minimum temperature of 7° C. (45° F.) in sun or shade.

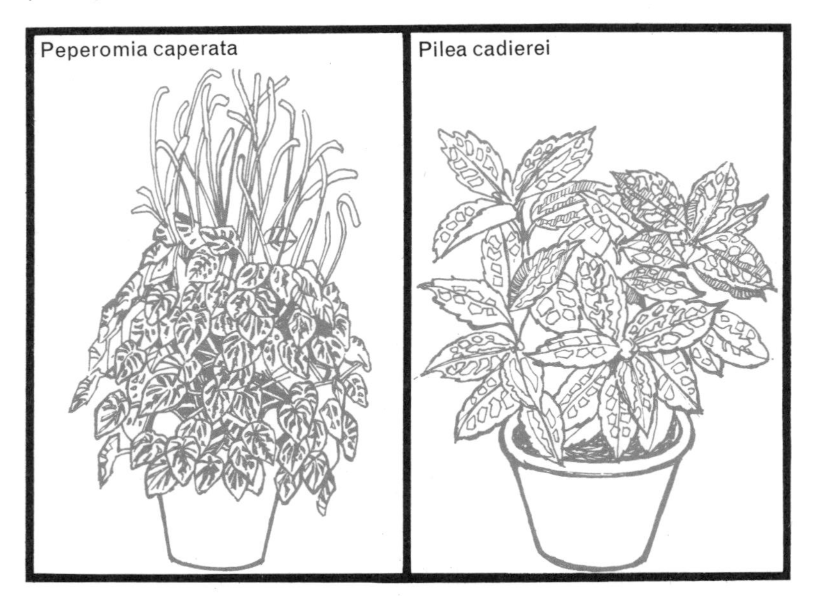

Pineapple and Others

The variegated pineapple, *Ananas comosus variegatus*, is a most effective plant for the house but it likes warm conditions and must not be exposed to draughts

It is the variegated pineapple, *Ananas comosus variegatus*, that is grown as a house plant. When in good condition it is very handsome, with rosettes of narrow, serrated, cream-edged leaves. Unfortunately these leaves easily spot if water lodges on them so the plant should be watered carefully. Do not let plants stand in saucers of water, the soil should be nicely moist throughout, no more. Grow it in J.I.P. No. 2 compost mixed with an equal bulk of peat. Maintain a minimum temperature of 15° C. (60° F.) and keep in a light window. Pineapples grown in rooms are unlikely to fruit, but if grown in a warm greenhouse they should fruit after about four years. Plants are increased either by detaching offsets (slips) from the base when re-potting in spring or by cutting off the top rosette of leaves from the fruit and rooting this in a warm propagator.

Any of the oranges, lemons or grapefruits (citrus) can be grown as pot plants, but the best for the purpose is the Calamondin Orange, *Citrus mitis*, because of its compact habit and ability to fruit when quite small. Its cultivation is described on page 71. Indoors put it in a well-lighted window and, if possible, stand the plant outdoors in a warm sunny place from June to September.

The Avocado Pear, *Persea gratissima*, described on page 71, makes a handsome house plant if the tips of the shoots are pinched occasionally to make it branch.

Sansevieria, Saxifraga

The fleshy, strap-shaped leaves of the sansevieria form an effective background and they contrast well with the other house plants in this easily imitated arrangement

There are several different kinds of sansevieria but by far the most popular is *Sansevieria trifasciata laurentii*, known as Mother-in-law's Tongue. It has fleshy, strap-shaped, slightly corkscrewed leaves banded and striped with yellow. It is an extremely hard-wearing plant which will grow in sun or shade and will survive in a room with a minimum temperature of 7° C. (45° F.) though it prefers a little more. It should be watered rather sparingly at all times and should be fed monthly from May to August.

The saxifrage that is grown as a room plant is *Saxifraga sarmentosa*, also known as *S. stolonifera*, popularly called Mother of Thousands because of the slender runners it produces, each carrying numerous plantlets which can easily be rooted if in contact with soil. It has sprays of small pink flowers and there is an attractive variety named *tricolor* in which the leaves are splashed with cream and rose. These plants should be grown in fairly light places, near a window. They are virtually hardy, do not need artificial heat and should be watered freely in spring and summer, moderately in autumn and winter.

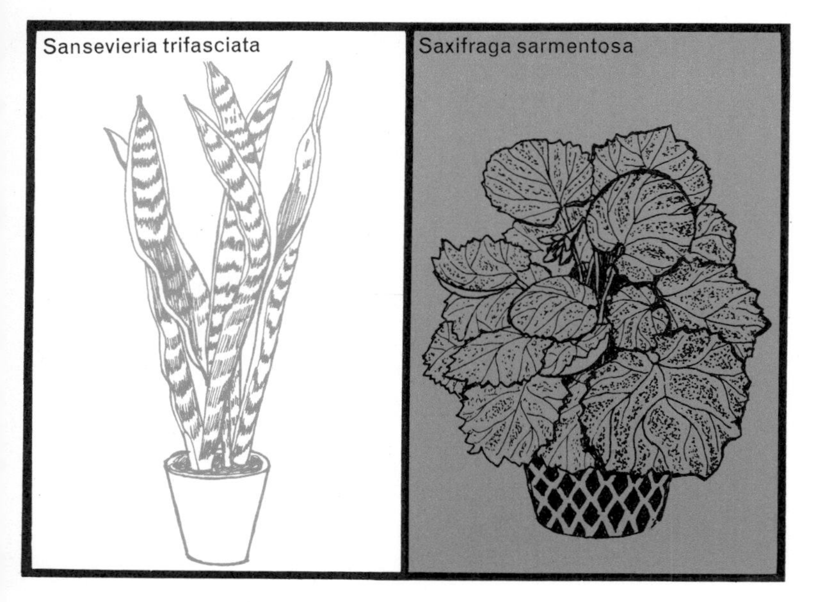

Sansevieria trifasciata

Saxifraga sarmentosa

Schefflera and Others

Elettaria cardamomum is a native of India and is one of the chief sources of cardamom. This variegated form makes an especially attractive foliage plant

Schefflera actinophylla (also known as *Brassaia actinophylla*) is an evergreen tree which can reach a height of 10 ft. or more in a tub, but it is slow growing and can be kept for years at much more reasonable dimensions in a 6- or 7-in. pot. It makes a single stem with long leaf stalks carrying leaves composed of numerous large green leaflets arranged in a fan. Grow it in J.I.P. No. 2 compost in a light place but not in direct sunshine. Water normally and maintain a temperature above 10° C. (50° F.). Increase by seed sown in a warm propagator in spring.

Elettaria cardamomum is a herbaceous plant with firm stems and broad dark green leaves which are cinnamon scented. It can be grown in J.I.P. No. 1 or peat-based potting compost and does not mind a considerable degree of shade; it should always be kept away from direct sunshine. Water fairly freely from spring to autumn but very sparingly in winter. Increase by division when re-potting in spring.

Almost any kind of pittosporum can be grown as a house plant, but the best for the purpose are *Pittosporum eugenioides* and its variegated variety and *P. tobira*. All are shrubs or trees grown primarily for their shining evergreen leaves, but *P. tobira* also produces clusters of scented white flowers. They are easily grown in pots in the same way as schefflera but are hardier. Increase by summer cuttings in a propagator.

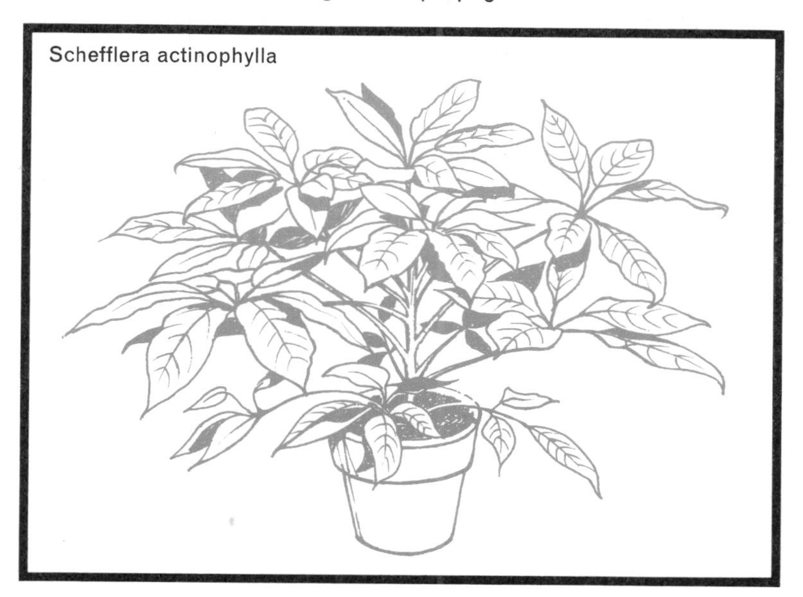

Schefflera actinophylla

Cacti and Succulents

Succulents are plants with very fleshy leaves or stems which enable them to store considerable quantities of water and so survive periods of drought. Many of them are natives of hot countries and must be grown in greenhouses or well-lighted rooms in countries where frosts are severe or prolonged, but some, such as many sempervivums and sedums, are quite hardy.

Cacti are a family of succulents most of which have no leaves, or only minute leaves, their function being taken over by the swollen body of the plant. The shapes of these plants are often very strange; they are frequently spiny and the flowers are borne on the main body of the plant. In some, especially the epiphyllums and some allied kinds, these flowers are very showy. Most cacti are easy to grow but many are too large to be conveniently accommodated except as young plants. Since all are accustomed to dry air many make excellent room plants if they can be given sufficient light.

Succulents other than cacti are so numerous and varied that few generalisations can be made about them. Most are as easy to grow as cacti.

It is the enormous variety to be found in cacti and other succulents that is one of their great attractions. Large collections can be built up of many different genera and species from all parts of the world. Because of this wide distribution in nature, cacti and succulents differ greatly in the temperature range to which they are accustomed, some being completely hardy and some injured by the slightest frost. Yet in this respect they are exceedingly adaptable and there is certainly no need to group them into different temperature groups as recommended for other greenhouse plants. Most cacti will thrive in any place with a minimum winter temperature of 4° C. (40° F.) and even the more tender succulents are quite content with a minimum of 10° C. (50° F.) provided they are kept fairly dry in winter.

Since they are able to live without water for quite long periods, cacti and succulents can be very convenient plants for gardeners who have to be away from home a great deal. For a similar reason they do not present any special problems when short holidays are taken. All the same it is a mistake to think of them as plants which need to be watered sparingly at all times. Most come from places where periods of drought alternate with rainy seasons when they are able to replenish their reserves, make their growth and complete their flowering.

Cacti and succulents can be grown in pots and pans like other greenhouse plants or they can be planted in little indoor gardens made with beds of suitably gritty soil, perhaps worked into irregular mounds with a few stones half buried in the soil. Such gardens should not be overplanted since cacti and succulents do not look their best when huddled together. Much of their charm lies in their highly individual shapes and they must always have sufficient space for these to be appreciated clearly.

Care of Plants

Aporocactus flagelliformis bears the imaginative common name of the Rat-tailed Cactus. As with all cacti that are grown in the home it needs careful watering

Contrary to popular belief, cacti and succulents should not be kept dry for most of the time. While they are growing they may need watering as frequently as non-succulent plants since the porous soil in which they are grown tends to dry out quickly. From April to September examine the plants daily and water any that appear dry. Give sufficient water so that it soaks right through the soil and starts to trickle out at the bottom, then do not re-water until the soil is dry again. From October to March it will be enough to examine the plants once or twice a week. Apply the water direct to the soil, and not to the plants. A few kinds, notably the stone mimics (page 29), resent any water lodging in them and are better watered by standing the pots for a few minutes in water almost up to their rims.

No artificial heat is usually required from April to October, but in winter frost must be excluded for all tender kinds. A minimum winter temperature of 7°C. (45°F.) suits most tender cacti and succulents. High temperatures in winter are undesirable.

Most cacti and succulents enjoy sunshine, but some prefer a degree of shade in summer. This is specially true of the epiphyllums and their allies, and also of haworthias and gasterias.

Potting and Handling

This varied collection of cacti and succulents provides an unusual decorative feature for the home. The container used must have ample outlet for surplus water

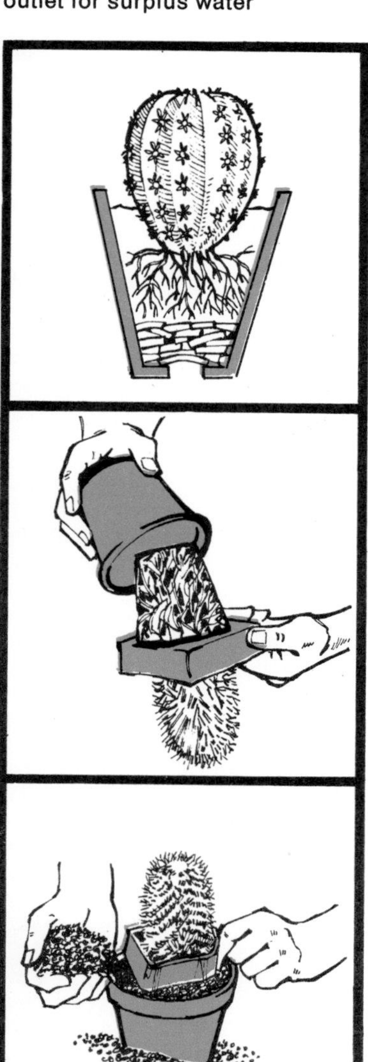

Cacti and succulents can be grown in any reasonably good soil that is sufficiently porous. John Innes No. 2 Potting Compost with the addition of 1/6th its own bulk of very coarse or sharp grit suits most kinds of cacti and succulents, and they can be grown in ordinary clay or plastic flower pots or in any container that has ample outlet for surplus water.

Pot cacti and succulents in March or April or when they are just starting to grow. Place some pieces of broken clay pot or a few small pebbles or stone chips in the bottom of the container to improve drainage, and, without disturbing the roots, place the plant in the pot and carefully work the new soil around the old.

Some cacti are so spiny that it is difficult to hold them. One way of dealing with the problem is to take a sheet of paper and fold it several times to form a thick band. Wrap this round the body of the plant and grip the two ends so that the paper holds the plant without the fingers coming into contact with it.

Take care not to break off the spines when handling cacti as they will not grow again and the damage spoils the appearance of the plant.

Increasing Plants

The enormous and beautiful flowers of epiphyllums appear from April to June. Like most cacti and other succulents, they are very easily propagated from cuttings

Many cacti and succulents can be increased by division when repotting. Some make numerous offsets which can be detached easily and even if they have no roots they will generally form them readily if placed in very gritty soil and kept moist.

Many succulents can be grown from individual leaves treated like cuttings and pressed into gritty soil. Some varieties do this naturally, the leaves falling off and rooting where they lie on the damp soil. Bryophyllums, for instance, form plantlets on the edges of the leaves, which fall off and subsequently root and grow.

Branching succulents can usually be increased by inserting short pieces of stem as cuttings and with some cacti which do not make offsets, pieces of the plant can be severed and treated as cuttings.

Many cacti and succulents can also be grown from seed and some firms offer packets of mixed seeds which provide a cheap and interesting way of starting a collection. Sow the seeds in February or March in John Innes Seed Compost, do not cover them at all or at the most sprinkle coarse sand over them and germinate at 21°C. (70°F.). Allow the seedlings to make tufts of roots before pricking them out, using similar compost, and later pot them singly.

Keeping Plants Healthy

The genus *Notocactus* provides many easily grown plants with striking flowers. A watch should be kept for pests and diseases but these are not usually a major problem

Cacti and succulents are not, as a rule, greatly troubled by pests or diseases.

Mealy bug, a small bug-like creature protected by a covering of a white waxy substance, may attack the plants above or below ground. Above the surface it can be seen and picked off with a pointed stick or the plants can be sprayed with malathion, but below ground it may well be overlooked. Watch for it when re-potting and, if present, wash it off with malathion.

Scale insects attach themselves to plants like minute limpets but they too can be killed with malathion.

Red spider mites are so tiny that they can scarcely be seen without a hand lens, but small brown spots towards the tips of plants are a warning to look out for them. If present, spray with malathion or fumigate with azobenzene.

Slugs and snails can be controlled with metaldehyde bait, woodlice and ants by dusting with an insecticide, and mice can be trapped or poisoned.

Rot may be due to various fungi encouraged usually by overwatering. Cut out all the infected growth, dust the area with flowers of sulphur or quintozene and keep the plant rather dry for a time.

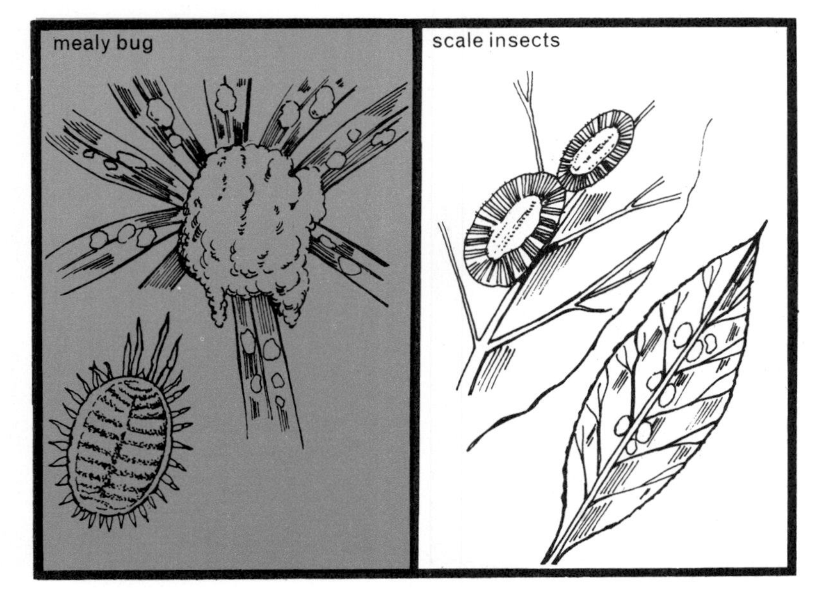

Small Cacti

Rebutias are dwarf-growing cacti which flower readily. They are very easy to grow and when properly cared for they may even produce flowers as one-year-old seedlings

Aporocactus flagelliformis: the Rat-tailed Cactus, so called from its long, narrow, trailing stems. The flowers are cerise.

Astrophytum: known as the star cacti, the plants of this group are globular with star-like ribs and red and yellow flowers on top.

Chamaecereus: clusters of small, spiny, cylindrical stems with scarlet flowers in early summer.

Echinocereus: a varied group, some upright, some sprawling, with attractive flowers in many colours. Even tiny plants will flower.

Echinopsis: barrel-shaped plants markedly ribbed with white or creamy flowers on top. There are numerous kinds.

Gymnocalycium: small globular plants with quite showy flowers on top. All kinds like some shade in summer and free circulation of air.

Lobivia: usually globular or cylindrical plants which flower while still quite young. There are many kinds with flowers ranging from golden yellow and orange to pink and carmine.

Mammillaria: globular or cylindrical plants with spines and some with wool also. Small flowers are borne in circles around the tops of the plants ranging from white and yellow to pink, crimson and green. There are more than 200 kinds.

Notocactus: globular, usually flat-topped plants, ribbed, with red and yellow flowers on top. They are very easily grown.

Rebutia: almost globular plants with flowers around the bottom in yellow, salmon, pink and carmine.

Large Cacti

The golden spines and globular shape of *Echinocactus grusonii* give it the common name of the Golden Barrel Cactus. It can grow to nearly 3 ft. across

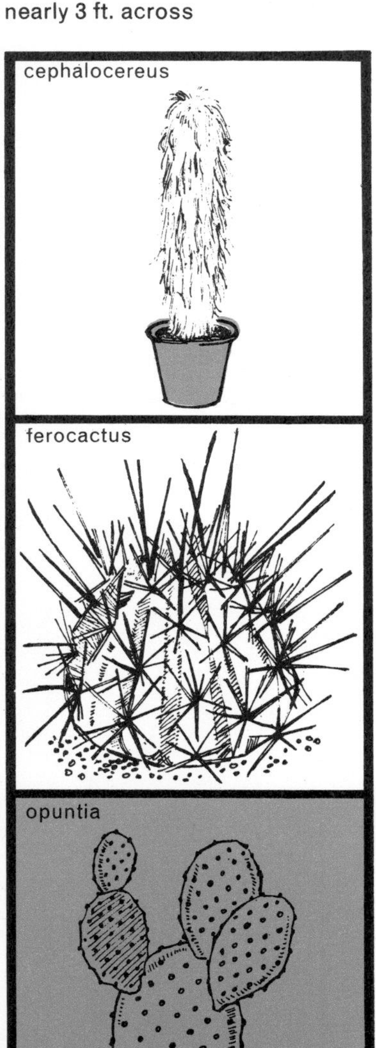

Cephalocereus: tall cylinders covered in long hairs. This characteristic is most marked in *C. senilis*, known as the Old Man Cactus, a great favourite.

Cereus: tall columns of growth with fine flowers, usually white, which open at night. Only quite old plants flower.

Cleistocactus: tall columns with small red flowers on the side.

Echinocactus grusonii: a globular-shaped cactus with golden spines known as the Golden Barrel Cactus or Mother-in-law's Chair.

Ferocactus: a group known as the barrel cacti because of their shape. All kinds have long, often brightly coloured spines.

Harrisia: plants with long thin growths which sprawl about unless tied to some support. The white flowers, which are scented, open at night.

Lemaireocereus: allied to cereus, with smooth, ribbed stems. This group needs warmth and light.

Opuntia: large circular, flat 'pads' built one on top of another which often produce very large plants. The Prickly Pear, a troublesome weed in some hot countries, is an opuntia.

Trichocereus: another group allied to cereus and easy to grow. They make long, branching plants with showy white flowers, but these are likely to be produced only by mature plants.

Epiphytic Cacti

Schlumbergera buckleyi, the Christmas Cactus, was formerly known as zygocactus and may still be listed under that name. It is a very popular plant and is widely grown

An epiphyte is a plant which receives most of its food from the air. In nature such plants often live in the branches of trees or on damp rocks. A number of cacti have these characteristics and nearly all of them are remarkable for the size and beauty of their flowers. In summer most of these cacti prefer a certain amount of shade and from June to September they can be grown in a shelter made of lath slats allowing the free circulation of air, or suspended from a small tree.

The principal kinds are epiphyllum, schlumbergera (zygocactus) and rhipsalidopsis. There are many hybrids in various colours including shades of pink and carmine.

Schlumbergera buckleyi (formerly zygocactus) is known as the Christmas Cactus because it starts to flower in December, continuing until February. The flattened stems have a graceful arching habit. *Rhipsalidopsis gaertneri* (formerly *Schlumbergera gaertneri*) has a similar habit but flowers in March and April. It is known as the Easter Cactus.

Epiphyllums themselves are usually stiffer and more erect in habit. They flower from April to June and rest in December and January, at which period they require little or no water. From March to June water fairly freely, sparingly for a few weeks after that, then more freely as they start into growth again.

Agave, Aloe

Agave americana is chiefly grown for its large rosettes of striking foliage. There are several varieties with banded or striped variegations, this one being *Agave a. marginata*

These plants with stiff, fleshy leaves, usually arranged in rosettes, are all easy to manage, though some are too large to be conveniently grown in pots except while young. They should be watered fairly freely in spring and summer, moderately in autumn and winter and grown in a light, sunny place.

Agave americana is known as the Century Plant because it is so slow in coming into flower. The very tall spikes are striking rather than beautiful and the plant is grown for its large, blue-grey, spine-tipped leaves which in some varieties are striped or banded with cream or white. It can be grown out of doors in some mild seaside gardens and can be planted out from May to October almost everywhere. Other smaller kinds are *Agave filifera*, with threads along the edges of the leaves, and *A. victoriae-reginae*, with dark green leaves decorated with white lines and hairs.

The most popular aloe is *A. variegata*, a neat plant known as the Partridge Aloe because of the grey and white markings on the leaves. So is *A. plicatilis*, the Fan Aloe, so called because the stiff, blue-grey leaves are arranged in a fan, not a rosette. *A. striata*, the Coral Aloe, has leaves edged with pink.

Gasteria and haworthia are allied to the aloe, but enjoy more shade and can be used as house plants. They are small plants with tapered leaves, often striped or spotted with white like the Partridge Aloe.

Agave victoriae-reginae

Aloe variegata

Crassula and Others

Kalanchoes are often sold by florists and greengrocers as house plants at Christmas. They should be watered carefully and kept at a minimum of 13° C. (55° F.) in winter

Crassulas, kalanchoes and rocheas are closely related plants with fleshy leaves which are sometimes confused with one another. The most useful as pot plants are *Crassula falcata*, with fleshy grey leaves and scarlet flowers in summer; *Kalanchoe blossfeldiana* with little sprays of scarlet or yellow flowers in winter, and *Rochea coccinea* (it is sometimes called *Crassula coccinea*) with close-packed leaves and tight clusters of scarlet flowers in summer. There are many other good crassulas, including *C. arborescens* which in time makes a big plant, *C. lycopodioides* and *C. tetragona*, both like tiny trees.

Pot these plants in spring in the smallest pots that will accommodate the roots comfortably. Water them fairly freely in spring and summer; sparingly in autumn and winter in the case of crassula and rochea, but more generously for kalanchoe which is then in its growing period. Keep them in a light, sunny place without shading and with a minimum winter temperature of 7° C. (45° F.), though 13° C. (55° F.) will be better for kalanchoe as it will then be in flower.

Raise kalanchoe from seed sown in March or April for winter flowering, prick off into boxes and later pot singly in 3-in. pots.

Raise crassula and rochea from cuttings of firm young shoots in very sandy soil in May, June or July.

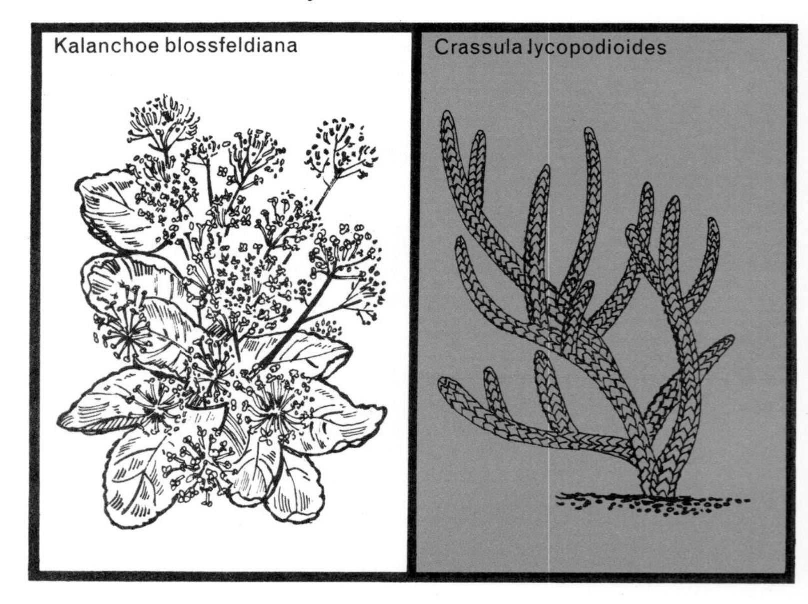

Kalanchoe blossfeldiana

Crassula lycopodioides

Euphorbia

There are lots of succulent euphorbias and many of these differ greatly in appearance. This one, *Euphorbia obesa*, needs very good drainage and little or no water in winter

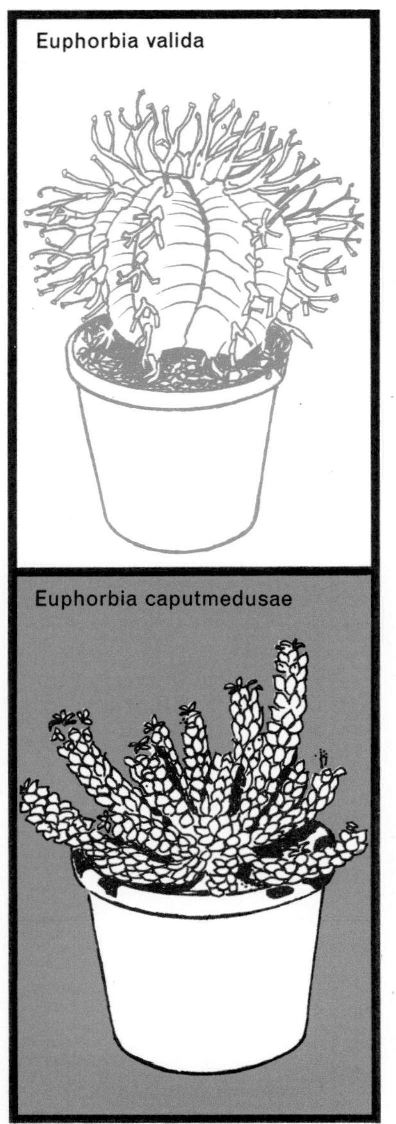

Not all euphorbias are succulents. Some are hardy herbaceous plants and one, the Poinsettia (*Euphorbia pulcherrima*) is a popular greenhouse shrub grown for its scarlet bracts which appear in winter. But there are also many succulent kinds of euphorbia, many of which differ greatly in appearance. Some, such as *Euphorbia abyssinica* and *E. grandidens,* are branched and tree-like, others such as *E. clava* and *E. clandestina* have a single stem with leaves on top like a little palm tree. Yet others, such as *E. meloformis*, *E. obesa* and *E. valida*, have very swollen stems like cacti. Then there is *E. caputmedusae*, which means Medusa's Head, an allusion to the grey-green snake-like branches which radiate from a ball-like central stem. Very popular, too, are those with vermilion bracts, like flowers, such as *E. splendens* and *E. bojeri*, both of which are known as Crown of Thorns or Christ's Thorn, because they are so spiny.

Even the largest of the euphorbias can be grown in pots when young. Give them a minimum temperature of 13° C. (55° F.) and water most of them moderately in spring and summer, sparingly in autumn and winter, but hardly at all for *E. splendens* and *E. bojeri* in January and February. Prune these two in May to keep them bushy.

Cultivation of euphorbias under greenhouse conditions is given on pages 76 and 77.

Lampranthus

Lampranthus roseus has cheerful pink flowers and is one of the taller-growing kinds in this group. It can be planted out in a sunny border from June to September

Most of the plants which are commonly called mesembryanthemum have been renamed lampranthus by botanists. Most are sprawling plants with small, fleshy leaves, but some make stiffer, more bushy growth. Many have showy flowers opening fully only in bright light. They are easily grown in J.I.P. No. 1 compost in a cool greenhouse without shade, watered moderately in spring and summer, then sparingly.

Water all kinds freely in spring and summer, sparingly in autumn and hardly at all in winter. Cuttings of firm young stems root readily in summer and make good flowering plants the following year.

Among the showiest kinds are: *Lampranthus aurantiacus*, bushy to 1½ ft., orange-red flowers; *L. blandus*, similar, but pink flowers; *L. brownii*, to 1 ft., smaller orange-red flowers; *L. coccineus*, sprawling scarlet flowers; *L. falcatus*, sprawling, pink, scented flowers; *L. roseus*, bushy to 2 ft., pale pink.

Cryophytum crystallinum is closely allied but is grown for the decorative effect of the shining watery spots which cover its leaves and prostrate stems, giving it the popular name Ice Plant. It is an annual readily raised from seed sown in a temperature of 15 to 18° C. (60 to 65° F.) in March and grown on as for Lampranthus. Similar culture suits *Dorotheanthus bellidiflorus* (often called *Mesembryanthemum criniflorum*), the Livingstone Daisy.

Sempervivum and Allies

There are numerous species and varieties of sempervivum, the popular houseleek. These massed rosettes belong to *Sempervivum tectorum calcareum* Mrs Guiseppe

Sempervivum Commander Hay

The true sempervivums or houseleeks are hardy and can be grown out of doors successfully provided they do not get too wet, especially in winter when they can easily rot away. But their near relatives, the aeoniums and aichrysons, are tender except in very mild places. All are excellent plants for growing in sunny greenhouses in pots or pans, the hardy kinds with just sufficient heat to keep out frost. All enjoy a very freely drained, fairly rich soil with plenty of coarse grit or sand. J.I.P. No. 2 compost with the addition of half its bulk of extra sand will suit them well and pots or pans can be filled to a quarter of their depth with broken crocks or gravel to ensure that soil is never sodden. They all enjoy full exposure to light and sunshine at all times as this improves the colour of the leaves, and they should also be ventilated freely, subject to the necessity to protect species of aeonium and aichryson from frost.

Species and varieties of sempervivum are numerous and names are often confused. They range from the tiny, filament-covered rosettes of *Sempervivum arachnoideum minimum* to the 6-in. plum-purple rosettes of Commander Hay. The aeoniums are much bigger, sometimes branched and almost shrub like. *Aichryson domesticum* has much smaller leaves heavily variegated with white in some forms. It is freely branched and is an excellent house plant, which will put up with some shade.

Stapelia and Allies

The exotic flowers of the stapelias have an attractive star-like shape and are handsomely marked. In some species, however, they have an unpleasant smell

Stapelias have fleshy stems, ribbed and notched, but it is the strangely marked flowers which make them so attractive. These are shaped like a star-fish and are striped, speckled and mottled in the oddest way with brown, purple, maroon and yellow. Some, but not all, have an unpleasant smell and stapelias are sometimes called the carrion flower. There are many different kinds but the best known is *Stapelia variegata* and its varieties with yellowish flowers, variously speckled with brown, maroon and crimson.

Allied to stapelia is caralluma, with more bell-shaped flowers; duvallia, a smaller plant; echidnopsis, stiffer and more branched; hoodia with cup-shaped flowers and tavaresia, whose stems are covered with bristly hairs.

All these plants should be grown in soil with a little more coarse sand or grit than recommended for most succulents. Water them freely in spring and summer, sparingly in autumn and winter, especially hoodia and tavaresia, and maintain a minimum winter temperature of 10° C. (50° F.). Re-pot them each spring and keep up a succession of young plants by severing the stems in summer and rooting them as cuttings.

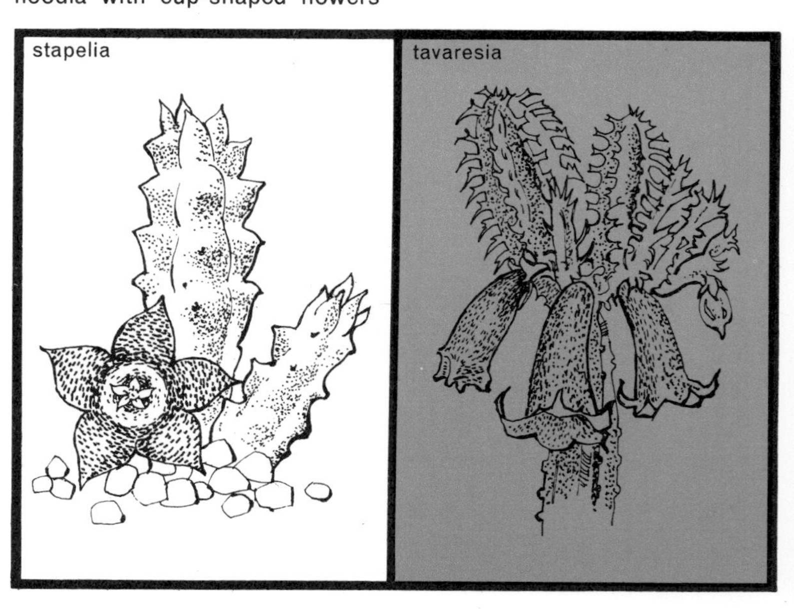

stapelia | tavaresia

The Stone Mimics

The stone mimics are very interesting subjects to grow. Lithops are the most pebble like of all and there are many kinds, differing in their colour and markings

Some succulents protect themselves in the wild by contriving to look almost exactly like the stones among which they grow. They are all small plants very suitable for cultivation in pots and though they may not be beautiful they are of great interest and variety.

Argyroderma consists of two segments like a pebble neatly sliced in half. The small flower is produced in the centre.

Conophytum has growths like fat buttons complete with a depression in the top from which the quite large flowers develop. There are many kinds, differing in shape, colour of growth and flower.

Fenestraria has a little transparent 'window' on top of each growth to let in the light, since in its natural state it grows embedded in sand with only the 'window' uncovered.

Gibbaeum has cleft growths like the mouths of some fish, for which reason it is called shark's head.

Lithops are the most pebble-like of all and are available in a variety of kinds, differing in their colour and markings.

Pleiospilos has pairs of leaves forming on top of one another but as the new leaves grow the old ones wither away.

All these plants like an extra ration of coarse sand or grit in the soil. Sink the pots in gravel or sand, give them the lightest possible position, and water them fairly freely from May to November, but very little for the rest of the year.

Other Kinds

Echeverias are handsome succulents. There are about 100 species and numerous hybrids and this one, *Echeveria retusa*, makes an excellent house plant which will grow to 2 ft.

The bryophyllums have leaves of various shapes and colours, with little plantlets on them which drop off and soon grow into new plants. Bryophyllums are now regarded as species of kalanchoe.

Cotyledons are branching plants with variously shaped leaves. One of the best is *C. orbiculata* with grey leaves and clusters of hanging yellow flowers.

Most echeverias make neat rosettes and the fleshy leaves are of various colours – pale green, blue-grey and bronze, sometimes marked with pink. *E. harmsii*, also known as *Cotyledon* and *Oliveranthus elegans*, is a branching plant with narrow grey leaves and large, scarlet and yellow flowers.

Kleinia articulata is known as the Candle Plant because of the shape of its stems, which are like old-fashioned tallow candles.

Many sedums (stonecrops) are hardy, but some are tender and make good pot plants. *S. morganianum* carries its silvery-green leaves in long tassels and can be suspended in a basket. *S. sieboldii* has blue-grey leaves with pale yellow centres in the variegated variety.

All these are easy plants to grow in J.I.P. No. 1 or equivalent compost in a cool greenhouse without shade or in a sunny window.

Echeverias are often used for bedding outdoors in summer, but in winter need to be watered with special care as the leaves may spot if water lies on them.

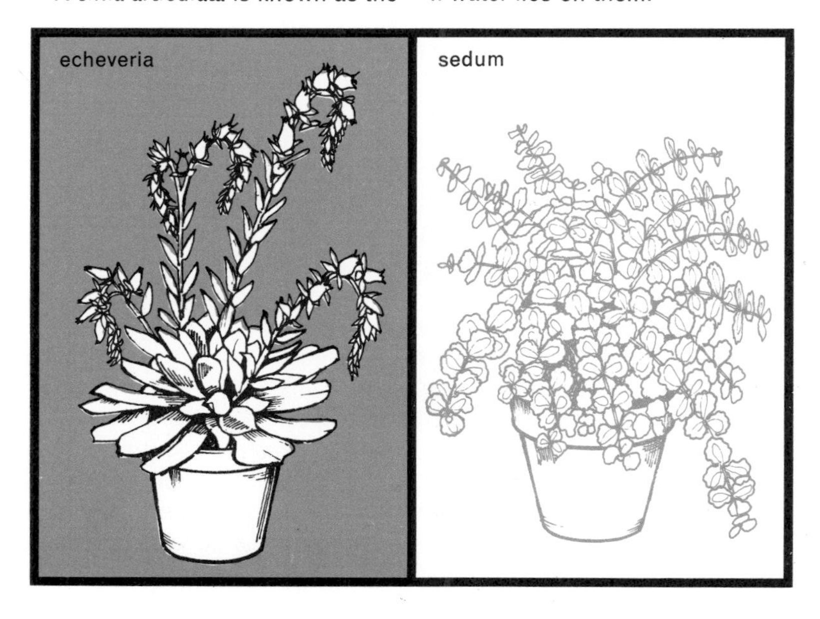

Bonsai

Bonsai is the Japanese name for dwarfed trees. These are quite different from topiary specimens, which are trees or shrubs clipped to artificial shapes. Bonsai reproduce the natural shape of each kind of tree, though they are usually modelled on ones which are very old or which have grown in exposed places and so developed romantic and interesting shapes. Bonsai may be single specimens or groups of small trees growing like a little grove or copse. They can be grown in ordinary flower pots but are usually put in special ornamental bowls made with proper drainage holes and small feet to raise the bowl off the surface, thus discouraging roots from forming a mat beneath the pot. The more light bonsai receive the better, as this helps to keep them dwarfed. They can be used for short periods for room decoration but should not be kept for long in poor light.

When purchasing ready formed bonsai specimens it is important to remember that the oldest plants are the most valuable provided they are well formed and in good health. This is in contrast with other kinds of nursery stock where old plants are usually to be avoided and it is young specimens with plenty of growth ahead of them that are most desirable. It is age which gives a bonsai tree its peculiar charm but very old specimens are inevitably rare and commercial producers of bonsai, not unnaturally, attempt to mimic the appearance of antiquity in young plants.

When choosing a bonsai tree examine it as follows. Its shape, however romantic, must be characteristic of its species and not so distorted as to look like some different kind of tree. The trunk must be substantial and well formed, tapering from base to top and not overburdened by its branches, which should be well spaced and must not arise so low down as to give the appearance of a bush. Some roots should appear on the surface, dipping naturally into the soil after a few inches but conveying the idea of an ancient tree with surface roots eroded by wind and rain. The soil must be kept high in the container so that the root formation is clearly visible viewed from the side. Neither roots nor branches should cross or rub, and the branches must not give a mop-headed appearance. A specimen representing a wind-blown tree, with most branches going one way, is better if it is set towards one end of a fairly long container so that the branches can extend over it, otherwise the effect will be unbalanced.

Many specimens have one specially attractive side, but all should be reasonably good viewed from any side. Late removal of a training wire may have chafed or ridged a stem, and removal of a large branch, which would not have been necessary had constant attention been given to the pinching of young shoots, can have left a large scar. Freedom from large scars or other blemishes is particularly important in deciduous specimens since the branches are so very open to inspection in winter.

Making Bonsai

Bonsai is the Japanese name for dwarfed trees grown, sometimes to a great age, in a handful of earth in a shallow container. A collection of these trees is shown above

Many useful specimens for bonsai can be found growing wild, particularly in rocky and exposed places. Look for seedlings that have been stunted or restricted by poor soil and lack of room to spread their roots, and have already acquired an interesting shape. Lift such plants with great care, preserving as much of the root system as possible, and pot them up at once. They will require shading and frequent overhead syringing for a few weeks while recovering from the shock of transplanting. Spring and autumn are the best seasons for moving such natural specimens.

Alternatively, bonsai can be grown from cuttings or seed, but seed is better as it is easier to train the seedlings in a natural manner. Sow the seed in spring in J.I.S. or peat seed compost, preferably in twos or threes in peat pots (to be singled later), and germinate them in an unheated greenhouse or frame.

If grown in peat pots it is easy to prune the roots when they penetrate the soft walls of the pots, and so growth can be restricted from the start. If grown in seed pans and transplanted singly to ordinary pots in John Innes No. 1 compost, it will be necessary to tap them out occasionally to cut off any coarse roots on the outside of the soil ball. As the seedlings grow some side buds should be rubbed out, so ensuring that branches only grow where needed.

Suitable Plants

This bonsai is *Juniperus sargentii*. Bonsai need the maximum amount of light to keep them dwarfed and they should be used as room decorations for only a short while

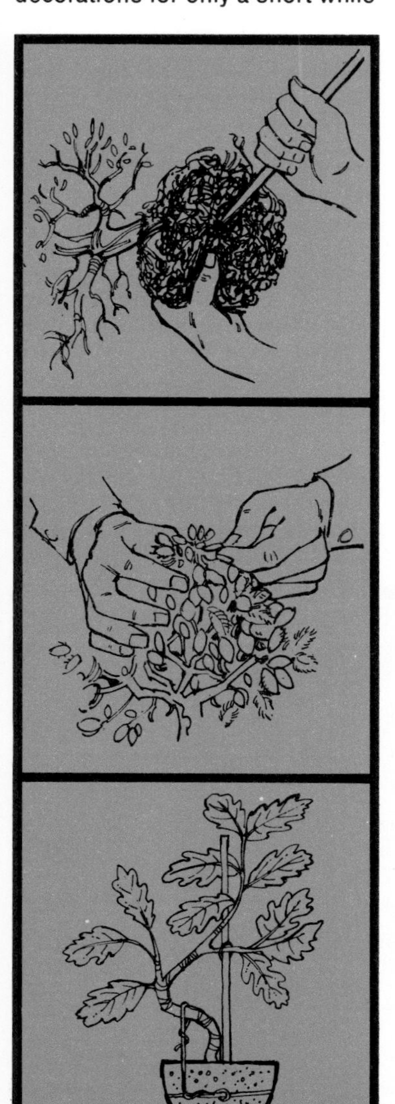

Bonsai must not be starved but be grown in J.I.P. No. 1 or equivalent. Vigorous kinds are re-potted annually, less vigorous ones every second or third year. When re-potting, tease out much of the old soil with a pointed stick and shorten or remove some of the coarser roots so that the tree can be returned to the same size container.

Start to restrict non-flowering bonsai just before they begin to grow each spring, when the growth buds likely to produce unwanted shoots should be carefully nipped out. Later, in spring and summer, shoots that are misplaced or those that are growing too big can be shortened or cut out where they join another stem. Prune all flowering bonsai when flowers fade but so as to keep the shape. Deciduous trees can be checked by carefully picking off all the young leaves. New ones will grow giving shorter, weaker growth and smaller leaves.

Keep the soil moist at all times, especially in spring and summer.

Young stems can be trained in any direction using soft fillis and temporary canes. Start with a slight pull and shorten the tie as the stem stiffens. Or use copper wire which can be progressively bent to force the stem into the required shape. Do not wire young or newly potted trees and keep constant watch that the bark is not chafed. Although the wire must be left on until the stem is woody, make sure it is not cutting into the bark as wounds are difficult to hide.

A distinctive habit and branch pattern are points to look for when choosing a small tree or shrub for bonsai. A low-growing cotoneaster is one of many suitable shrubs

Any small trees and some shrubs can be used for bonsai, but preference should be given to those with a distinctive habit and branch pattern. Pines of many different kinds are suitable and so are some species of spruce. Junipers are good; a favourite with Japanese bonsai artists is the naturally spreading, almost prostrate form of *Juniperus chinensis* known as *sargentii*. *Cryptomeria japonica*, cedars, larches and *Ginkgo biloba*, the Maidenhair Tree, are other good conifers for dwarfing.

Japanese maples are first favourites, but *Acer palmatum* may get die-back and may lose whole branches. Beeches and hornbeams both dwarf well, and zelkova is much used in Japan. Suitable flowering trees include plum, cherry, crab apple, hawthorn, Judas Tree and mulberry.

Fruiting trees should be chosen for the smallness of their fruit to maintain a sense of scale.

Good shrubs with which to form bonsai include dwarf forms of azaleas and rhododendrons, camellias, prostrate or low-growing cotoneasters, pomegranate (*Punica granatum*), pyracantha and various spindle trees. The bush form of ivy is also sometimes used and so are various hollies. Other possible evergreens are box and yew, but care must be taken to develop their natural habit and not merely to clip them into artificial shapes.

Greenhouse Management

Lack of a clearly defined management policy is often the main difficulty in greenhouses devoted to a variety of ornamental plants. Each plant has its own requirements and, though there may be a considerable measure of elasticity in these, it may be impossible to decide on any overall management policy for the house that will not spell disaster for some of its occupants.

It is sometimes suggested that differences of requirement can be met by segregating plants with like needs within the greenhouse. Some plants, we are told, should be grown at the warm end, others near the door where it is presumably cooler (and incidentally probably a lot draughtier) and so on. But the whole essence of good greenhouse management is to eliminate as far as possible variations of temperature within the house and to secure as even a distribution of air, whether warm or cool, wet or dry, as is possible.

It is wise, therefore, when choosing plants to eliminate any which have such very different requirements that they cannot be fitted in comfortably to an overall management routine. With this idea in view I have adopted in this book a system of greenhouse classification according to temperature, and have used this instead of repeating detailed instructions for temperature control throughout the year. Four basic regimes are considered as follows:

Cold greenhouse. No artificial heat used at any time of the year. In such a greenhouse winter temperatures may be little or nothing above those outside and in cold districts such houses are unsuitable for any plants that are injured by frost.

Cool greenhouse. In this, a minimum winter temperature of 7° C. (45° F.) is maintained with an average from autumn to spring of 10 to 13° C. (50 to 55° F.) rising in summer to 13 to 18° C. (55 to 65° F.).

Intermediate greenhouse. Minimum winter temperature of 13° C. (55° F.) with average autumn to spring temperatures of 13 to 16° C. (55 to 61° F.) rising in summer to 16 to 21° C. (61 to 70° F.).

Warm greenhouse. Minimum winter temperature of 18° C. (65° F.) with an average autumn to spring of 18 to 21° C. (65 to 70° F.) rising in summer to 21 to 27° C. (70 to 80° F.).

If plants are watered individually it is possible to vary the quantity for each according to its requirements at the moment. But in many modern greenhouses some system of automatic or semi-automatic watering is installed to save labour and then it does become necessary to keep plants with the same water requirements together. Fortunately a great many plants have similar needs, which may be briefly described as water applied freely from April to September while they are making their growth, but only moderately from October to March when they are resting. To avoid repetition I have called this 'normal watering' and have only given more detailed instructions where it is necessary to depart from this norm.

Types of Greenhouse

Spring is a very attractive time in the cool greenhouse when many brightly coloured plants can be grown. Included here are narcissi, cyclamen, primulas, veltheimias and cinerarias

Greenhouses may be either span-roofed or lean-to. In the former type the roof is usually, though not always, double-sided with an equal slope on each side, whereas the lean-to house has a one-sided roof. It is intended to stand against a wall, whereas the span-roofed greenhouse is free standing.

Either type of greenhouse may be glazed to ground level or to only about 2½ ft. above the ground with a wood, brick or concrete wall below that. Fully glazed houses are useful for plants grown in beds at ground level, such as shrubs and climbers; and also for tall plants such as greenhouse chrysanthemums. Greenhouses built on low walls are useful for pot plants that are to be grown on stages, also for raising seedlings and rooting cuttings where it is convenient to have plants at about waist level. The comparatively ill-lighted space under the staging can be used for storage, and also for shade loving plants such as *Begonia rex* and ferns.

Greenhouse framing may be of wood or metal. Wood is in general cheaper, but sash bars usually need to be rather thicker and so exclude some light. Western red cedar is much more rot resistant than deal and aluminium alloy framing is durable and requires little maintenance.

Either glass or plastic may be used for glazing, but some plastics deteriorate fairly rapidly, so they must be selected with care.

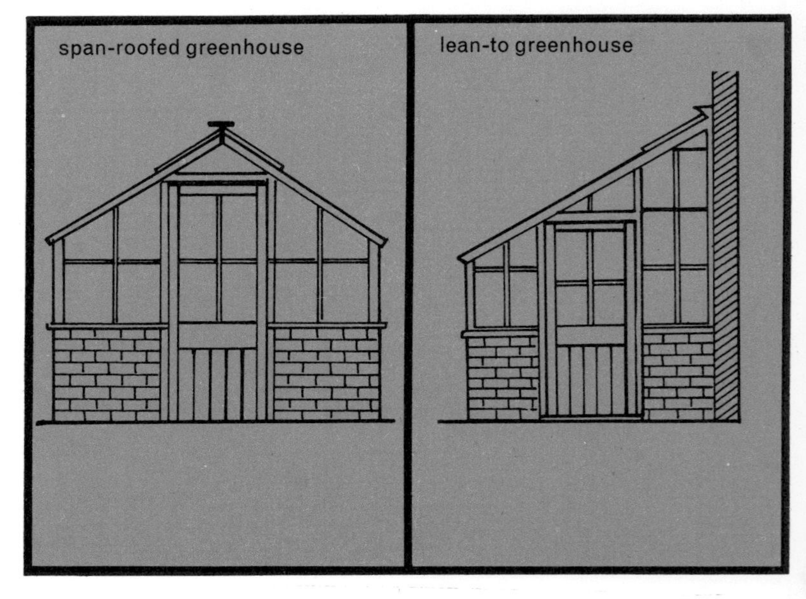

Choosing the Site

A conservatory extension to the house has a dual purpose as it is made for the use of both people and plants. Ideally, conservatories should be used for the display of plants when they are at their best

Even where space is limited it is often possible for a small lean-to greenhouse to be built. This type of house can often be heated by an extension of the home heating system

Occasionally greenhouses of other shapes may be preferred. Circular or octagonal greenhouses can fit compactly into some awkward places and the interiors can be very economical of space. Dome-shaped houses have been constructed and also tunnel-shaped houses which in section may be either semi-circular or have six or eight separate faces. Such houses let in a lot of light, but can be more costly to construct than the conventional types.

Commercial gardeners have been making increasing use of readily portable greenhouses covered with stout plastic sheeting. The basic forms are often tunnel shaped and the plastic usually needs biennial renewing.

Portability is generally of less interest to private gardeners and greenhouses covered with rigid plastic sheets, usually of the corrugated type, have a wider appeal, especially as they are easily constructed at home on a simple wooden frame.

Conservatories are intended primarily for the display of plants when they are at their best in flower, fruit or leaf. They are often attached to the house from which they may be entered directly by french windows or a door. As a rule conservatories are supplied from other greenhouses or frames in which the plants are grown to maturity.

With the exception of those to be used for ferns or other shade-loving plants, always place greenhouses in as sunny a position as possible. It is easy to exclude light, if necessary, by shading; much more difficult to provide extra illumination when light is lacking.

Place lean-to greenhouses against walls with a southerly aspect and span-roofed greenhouses in the open well away from the shade of trees, but if heating is to be installed do not overlook the possibility that a supply of hot water or electricity may be required from the dwelling house. If so it may be necessary to compromise and place the greenhouse near the main building even if this means some loss of light.

Make sure that the greenhouse is securely placed on substantial foundations and that it is level. It is important to select a spot where surface water is not likely to drain into it and to provide good access to the greenhouse in all weathers, preferably by means of a paved, gravelled or asphalted path.

If possible, arrange for a water supply inside the greenhouse. It is also convenient to have electric lighting, even if electricity is not used as a means of heating.

The situation of a conservatory is often determined by the room to which it is attached. Since flowering plants are unlikely to remain in it for long it is not essential for it to be in the sun.

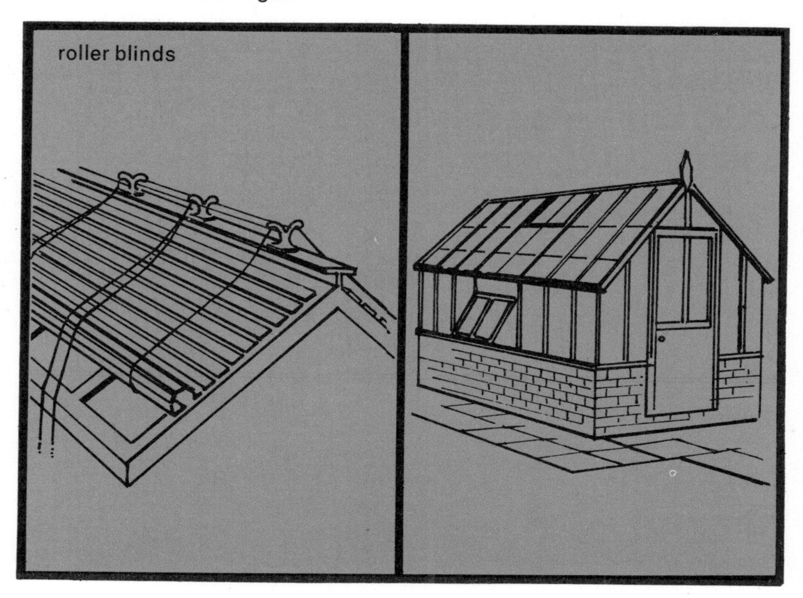

Heating

A paraffin oil heater designed specially for use in a greenhouse. Such heaters are simple and inexpensive to run, and greatly increase the range of plants that can be grown

Burning natural gas is an ideal way of heating greenhouses as it not only provides even, all-round heat but also is very economical to run. Portable heaters, as shown here, are readily available

Without some form of heating it will be impossible to exclude frost from the greenhouse at all times, and this will limit the range of plants that can be grown. Many of the most popular, useful or beautiful greenhouse plants are easily injured or killed by frost.

Heating may be by hot air, hot water, electric radiator, paraffin oil stove or gas. Hot-air heaters are frequently used in large greenhouses and can be economical and satisfactory, but they have not yet been much used in small greenhouses.

Hot water provides evenly distributed heat. The water may be heated by a solid-fuel boiler, a gas or oil-fired boiler or an electric immersion heater. All these are satisfactory, but solid-fuel boilers will need the most attention. If the dwelling house has central heating the possibility of extending the system to the greenhouse should be considered.

Ordinary electric radiators do not provide a sufficiently well-distributed source of heat. They must either be equipped with fans to blow the hot air around the greenhouse or be of the low-temperature type with a large radiating surface, such as tubular or panel heaters.

Oil stoves burning directly in the greenhouse are cheap to install and simple to run, but must be carefully cleaned and tended, as if they overburn, their fumes can be very damaging to plants.

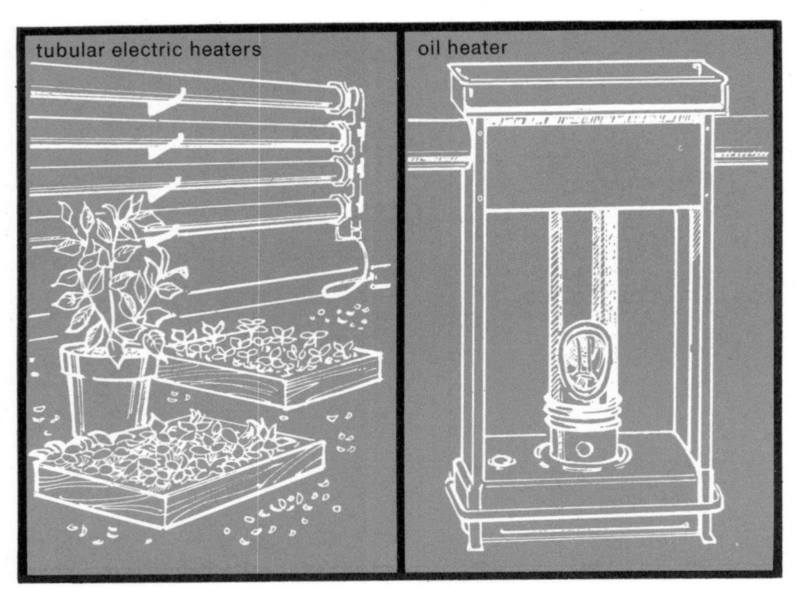

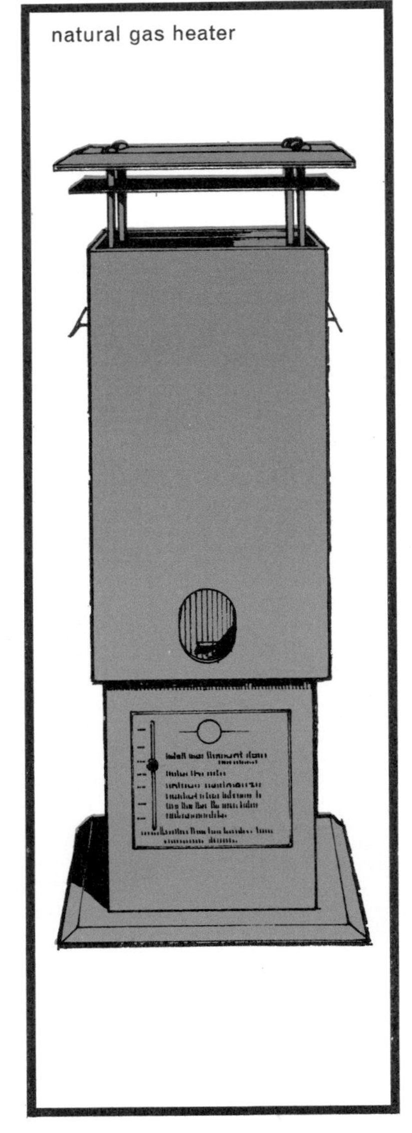

Boilers heated by coal gas can be used in place of solid-fuel or oil-fired boilers. However, care must be taken to site them so that there is no danger of fumes from the consumed gas entering the greenhouse as the sulphur compounds they contain are harmful to plants. This danger does not apply to natural gas as the by-products of burning this are carbon dioxide and water vapour, both beneficial to plants. In consequence free-standing stoves burning natural gas will allow the maximum use of heat with the minimum cost of installation.

North Sea gas is a natural gas of this type and stoves specially adapted for greenhouse use are available for use with it. There is no theoretical limit to the size of such stoves, but since the heat in a greenhouse needs to be distributed as evenly as possible, it is usually better to use several stoves spaced out in a large greenhouse rather than one big stove.

One of the most popular types is designed to give a 10° C. (50° F.) rise in temperature in a greenhouse of 100 sq. ft. area and in most parts of Britain one such stove would probably exclude frost from a house double this size. These stoves are thermostatically controlled and require a minimum of attention.

Ventilation

Good ventilation is extremely important. This greenhouse is well equipped with ridge ventilators and slatted blinds, so that the inside environment can be well controlled

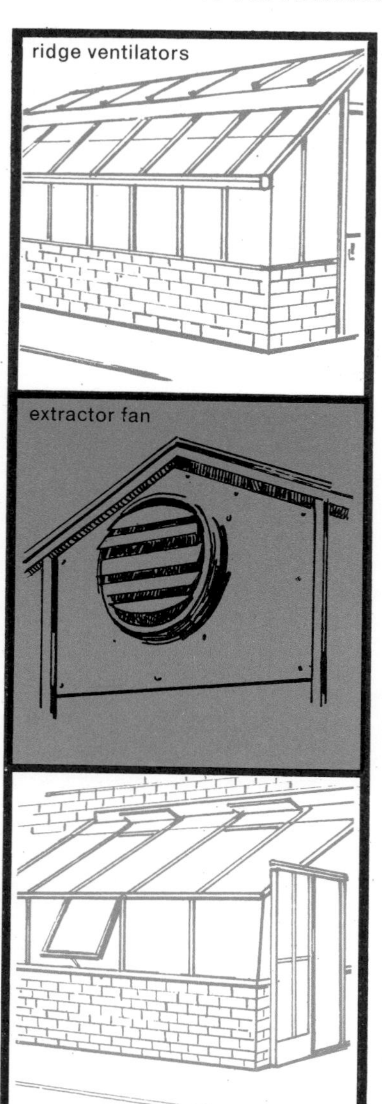

In sunny weather the air inside a greenhouse warms rapidly and can soon become too hot. Ventilation is essential to let this hot air escape and cooler air enter and, if required, to allow damp air to be replaced by drier air.

Ventilation may be effected by opening ventilators or by using extractor fans, both of which methods can be made automatic. The most important ventilators are those which are at or near the ridge of the greenhouse, since hot air rises and will escape most readily from the highest point. Ideally, ventilators should be equivalent to the length of the ridge and be placed half on one side, half on the other in a span-roofed house, so that they can be opened away from the prevailing wind. Strong draughts can be harmful to plants.

Extractor fans should be placed as high as possible, usually in the end panels of the greenhouse, and must be sufficiently powerful to change the air within the house every three or four minutes.

Ventilators in the sides may be useful in very hot weather to ensure a through current of air but they can create draughts if they are incorrectly used. The greenhouse door may also be left open on very hot days as additional ventilation.

The aim in ventilation should always be to maintain a reasonable temperature, a little higher during the day than at night.

Automation

Temperature control is a major factor in greenhouse management and many automatic devices are available to assist with this. Shown here is a thermostatically-controlled extractor fan

Greenhouses are affected by every change in the weather, small houses more rapidly than large ones because of the smaller volume of air they contain. Therefore anything that can be done to control automatically temperature, ventilation and humidity will greatly reduce the amount of time that must be devoted to the house.

Temperature control in winter depends mainly on controlling the supply of artificial heat. This can be done by a thermostat, which is now almost standard equipment incorporated with electric and natural gas heaters as well as oil-fired boilers and can be applied to some free-standing paraffin heaters. If a separate thermostat is used it should be placed as centrally as possible and be screened from direct sunshine.

In summer, temperature control depends mainly on ventilation and shading. Though various ingenious methods of automatic shading have been used none has yet caught on with the public. Automatic ventilation is a different matter since extractor fans can also be thermostatically controlled. Alternatively, ventilators can be automatically opened by a device operated by a very heat-sensitive fluid inside a cylinder. The cylinder contains a piston which is pushed outwards by the expanding fluid and in so doing lifts the ventilator.

Extra humidity is most likely to be required in summer and can be provided by a humidifier.

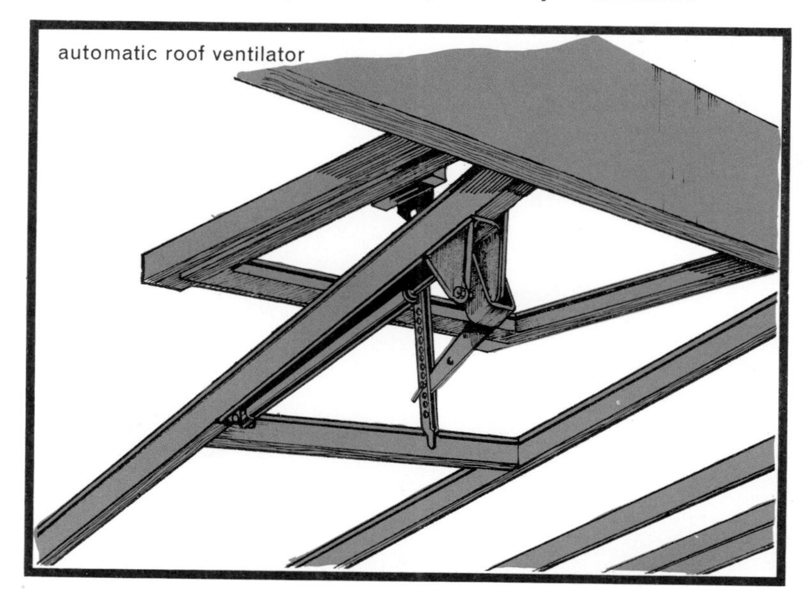

Watering

Plants in a greenhouse should be watered individually if possible. Care must be taken to direct the water at the soil, and not to splash the leaves or stems

Different plants require water in different quantities at different times of the year, three variables which make generalisation difficult. In general, the rule should be to keep soil moist throughout each pot or bed, but not to allow it to remain saturated for any length of time.

Provided soil is of the correct texture and drainage is adequate, surplus water will run away fairly quickly. So give sufficient water to soak the soil right through and then give no more until it begins to dry out. This can usually be decided by examining or feeling the surface of the soil, but when in doubt pick up a pot and feel its weight. Dry soil is lighter than wet and experience is soon gained in gauging the relative weights. In the case of beds, scrape away a little soil with the forefinger and see how moist it is an inch or so below the surface, where the roots are.

In spring and summer examine plants daily to see which ones require water and in autumn and winter examine them every third or fourth day. In very warm weather pot plants may require watering both morning and evening.

Apply water direct to the soil, not to the leaves or stems. Some plants like to have damp leaves, but these can be separately syringed overhead.

Automatic Watering

Some system of automatic watering, such as the capillary bench shown in the foreground above, is useful in the larger greenhouse or when the owner is absent for long periods

Various systems of automatic or semi-automatic watering are available, some electrically operated, others purely mechanical. One of the simplest is the capillary bench, which is a bed of sand or gravel kept constantly moist. Pot plants are placed on this and draw water from it by capillary attraction as the soil or peat they contain becomes drier than the bench. This bench must be quite level. It can be covered with asbestos cement sheets or polythene sheeting or the sand or gravel can be contained in shallow plastic trays. About an inch thickness of aggregate is required. Water can be supplied from a small tank fitted with a ball valve, by periodic trickle irrigation, or direct from the mains supply via a special control unit. Various proprietary kits are available.

Plastic pots without drainage crocks are best for use on capillary benches since they are thin and the soil inside them comes readily into contact with the moist sand or gravel on the bench. The capillary action must be started by watering each pot well after it has been placed in position.

Another system uses fine bore plastic pipes to convey water individually to each pot. Yet another uses a siphon tank to deliver water through rubber tubing fitted with drip nozzles, which can either be placed to discharge directly into the plant pots or on to a capillary bench.

Damping Down, Shading

Damping down a greenhouse with a watering-can fitted with a fine rose. This is an important job in the summer time, for it keeps the atmosphere of the house moist

In hot or dry weather gardeners spray or splash water about inside their greenhouses to evaporate and so cool the air and keep it moist. This is known as 'damping down' and is very important for some tropical and sub-tropical plants that originate from places with a humid atmosphere. It also helps to control red spider mites.

Damping down may be done with a syringe or with a watering-can fitted with a fine rose. Use clear water and thoroughly wet the floor of the house and the staging, especially if this is of a solid type covered with sand, gravel or ashes. Damp down in the morning and again at mid-day and in the evening too if conditions need it.

A hygrometer can be purchased to record the relative humidity of the air, but usually the look of the plants and the feel of the air on the skin is sufficient guide.

Shading is required to prevent direct sunshine from over-heating the air or damaging plants. It is more important for some plants than for others – hence the advantage of shading by means of roller blinds which can be readily raised or lowered as required. Alternatively, the glass can be sprayed or painted with lime-wash or a special shading compound. This is usually done in May and the shading is washed off at the end of September.

Staging and Beds

Gravel on greenhouse staging provides an attractive, moisture holding background for plants. Collect up leaves and other debris that fall on to the gravel or infection may be started

Plants may be grown either in pots, boxes or other containers or in beds of soil. If grown in containers it is usually most convenient to keep them on staging at about the level of an ordinary table, so that they can be examined and handled easily.

Staging may be either of open slat or of solid construction. Open staging is usually made of slats about 2 in. wide, spaced an inch apart. It can be made portable so that it can be removed quickly if the greenhouse is required for plants grown from floor level. Open-slat staging is best for those plants that require a fairly dry atmosphere, such as carnations and succulents.

Solid staging is often of asbestos cement sheets which rest on trestles and are covered with sand, small (pea) gravel, ashes or peat. These materials can be kept moist if desired and so this kind of staging is best for plants that appreciate a rather moist atmosphere, such as begonias and gloxinias.

Beds of soil for greenhouse plants can be constructed on solid staging, but are usually made at floor level. They should be at least 9 in. deep and may be separated from the natural soil with a sheet of polythene, but this should be slit or perforated to allow surplus water to escape.

Shelves slung from rafters or attached to side members can provide extra space and help to bring plants nearer to the light. Wire or plastic baskets can also be suspended above head level.

Soils and Composts

Collecting together the materials for John Innes compost. When making up compost, it is essential that all the ingredients are thoroughly mixed, so that plants receive a balanced amount of food

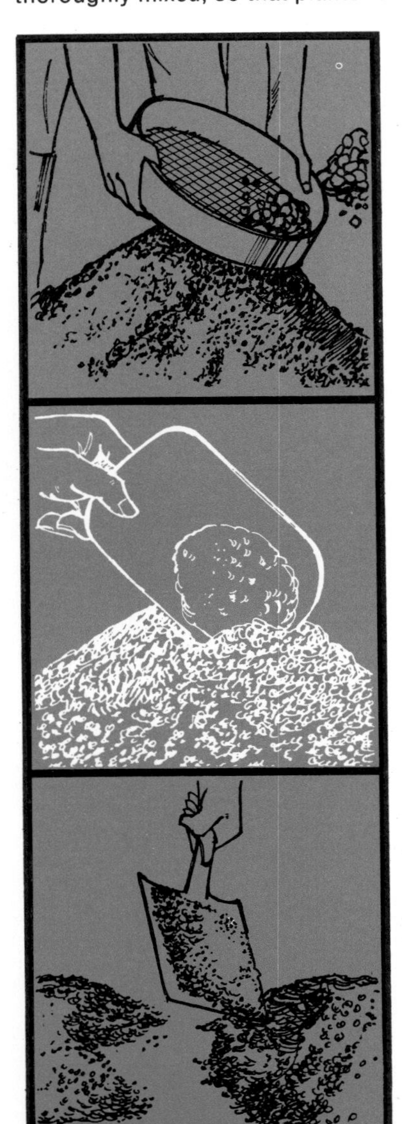

The soils or other growing media used for greenhouse plants are known as seed or potting composts. Standard composts will serve a wide variety of plants.

The most satisfactory soil composts are those known as the John Innes Composts.

The John Innes Seed Compost (J.I.S. for short) is prepared with 2 parts by bulk medium loam, 1 part granulated peat, 1 part coarse sand. To each bushel of this mixture add 1½ oz. superphosphate of lime and ¾ oz. ground limestone or chalk.

The John Innes Potting Compost is prepared by mixing 7 parts by bulk medium loam, 3 parts granulated peat, 2 parts coarse sand.

For the J.I.P. No. 1 Compost add to each bushel of this mixture ¾ oz. of ground limestone or chalk and 4 oz. of John Innes Base Fertiliser (2 parts by weight superphosphate of lime, 2 parts hoof and horn meal, 1 part sulphate of potash). For J.I.P. No. 2 use 8 oz. of J.I. Base Fertiliser and 1½ oz. of chalk or limestone per bushel of compost. For J.I.P. No. 3 use 12 oz. of J.I. Base Fertiliser and 2¼ oz. of chalk or limestone per bushel of compost. For J.I.P. No. 4 use 16 oz. of J.I. Base Fertiliser and 3 oz. of chalk or limestone per bushel.

The loam should be sterilised, since unsterilised soil may carry diseases. John Innes Seed and Potting Composts can be obtained ready mixed and so can several soil-less composts based on peat.

Pots and Containers

Flower pots are obtainable in a wide range of sizes. When potting up plants, one should be careful not to choose too large a pot in relation to the size of the plant

Greenhouse plants are grown in pots of various sizes and kinds and seedlings are usually raised in shallow boxes or trays or in special seed pans.

Pots and pans used always to be made of baked clay, but this is now being superseded by plastic which is lighter, easier to clean, and retains moisture and heat better. Seedboxes are usually made of wood and seed trays of plastic.

Pots are made in different sizes measured by the diameter across the top, which may be anything from 2 in. to 16 in. or thereabouts. For ordinary purposes, pots respectively 3, 4, 5, 6 and 8 in. across are the most useful. New clay pots are usually very dry and should be soaked in water before use, otherwise they will absorb water from the compost. All pots must be kept scrupulously clean and provided with at least one good hole in the base for drainage.

Wooden boxes must not be treated with creosote as the fumes from this substance are harmful to plants. Boxes can be used as they are or treated with a harmless preservative such as copper naphthenate. Tubs are often painted.

All boxes and trays must also have adequate holes or slits to allow surplus water to escape. Provided that suitably porous seed and potting composts are used, no extra drainage material will be required. Tubs are usually on short legs or castors. This allows water to drain away easily.

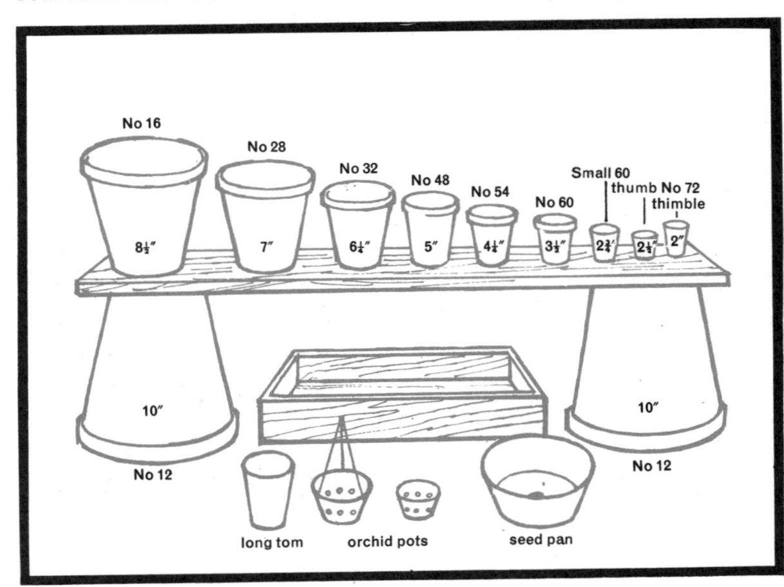

Potting

Potting up young plants in plastic pots. The compost is firmed lightly with the fingers, and is then watered thoroughly to settle it more securely around the roots

As a rule plants are moved on only one, or at the most two, sizes of pot at a time. Thus a plant in a 3-in. pot is re-potted into a 4- or 5-in. pot, then into a 6- or 7-in. pot and so on. A little plant in a large pot is seldom happy and it is very difficult to give it the correct amount of water.

Potting becomes necessary when the roots of developing plants have occupied all the soil in their pots and the plants are in danger of becoming starved. It can be done at practically any time of the year except winter.

To re-pot a plant turn it over, holding the pot in one hand and placing two fingers of the other hand over the mouth of the pot, one on each side of the plant. Rap the rim on something firm such as the edge of a potting bench, and the ball of soil should then slide out intact. Without disturbing the soil round the roots, place the plant on a little soil in the larger pot, trickle soil all round it and give the bottom of the pot a sharp rap on the bench to settle the soil in. With peat-based composts, nothing more is needed, but with soil composts a little firming all round with the fingers may be desirable. For very large pots a stick is occasionally used to ram the soil in round the edge.

Immediately after potting, water the compost well to settle it still more securely around the roots.

Feeding

Liquid feeding is especially useful for pot plants because the concentration of feed given can be so accurately controlled. Water dry soil before applying the liquid feed

The John Innes and various proprietary peat-based seed and potting composts all contain sufficient plant food to carry plants on for a considerable period, but there will come a time when supplies begin to run out and extra feeding is required. To judge the right moment requires some experience, but condition of growth is a guide. Feeding is most likely to be necessary between May and August and is hardly ever required in autumn and winter.

Either dry or liquid feeds can be purchased for greenhouse plants, and the dry feeds are divided into those which should be dissolved in water and those to be applied as they are. In all cases manufacturers' instructions regarding strength and method of application must be followed and should never be exceeded. It is better to give feeds that are too dilute than too concentrated.

A liquid feed containing minor as well as major foods can be made by suspending a bag of well-rotted manure in a tub of water and diluting the resulting liquor to pale straw colour. An alternative is seaweed extract used according to label instructions.

As a rule it is sufficient to apply liquid feeds about once every 7 to 10 days and dry feeds about once a fortnight. Dry feeds must be watered in and no more should be applied while any remains undissolved on the surface.

Sowing in Pots and Boxes

Primula obconica is a rewarding plant for the cool greenhouse. It can be raised from seed sown in March at a temperature of 15° C. (60° F.) and will then flower from December to March

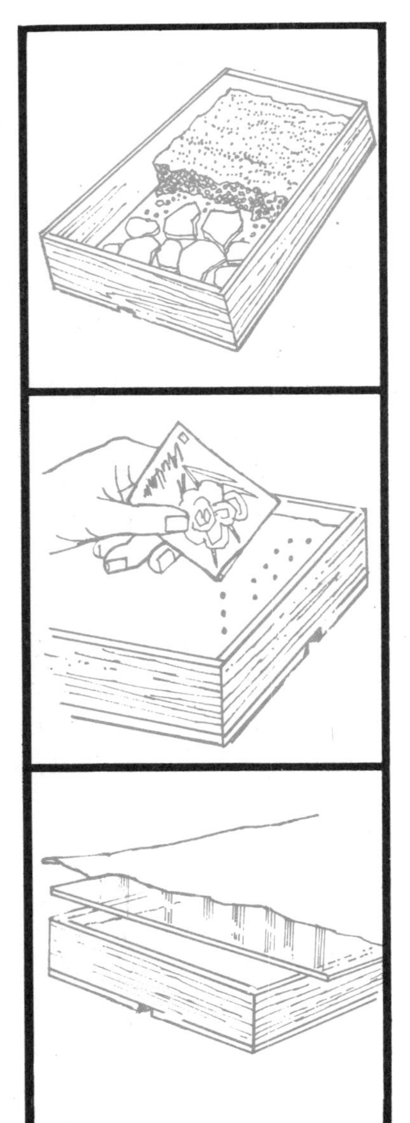

Pots, pans and boxes in which seeds are germinated must be well provided with holes or slits through which surplus water can drain away. Cover these outlets with pieces of broken pot (crocks), small gravel or coarse peat so that they cannot be blocked by fine soil and then fill up with seed compost. Press the compost in gently with the fingers and smooth it off level with a straight-edged piece of wood. Firm composts based on loam, such as John Innes Seed Compost, with a smooth wooden block, but do not firm peat composts. When ready for sowing, the surface of the compost should be about ½ in. below the rim of the receptacle.

Broadcast the seed thinly over the surface of the compost and cover by sifting soil, peat or sand over it. Very small seeds need not be covered at all, but both types should be protected by a pane of glass laid over the container (but not touching the soil) with a sheet of paper on top. The paper must be removed directly the seedlings appear and the glass a day or so later.

Water the seeds thoroughly using a watering-can fitted with a fine rose. For very small seeds, water by holding the pan for a few moments almost to its rim in a tub of water.

Germination and Watering

The sheets of glass used for covering seedboxes should be tilted slightly, to admit some air, once the seedlings appear and all the condensation wiped off daily

Most seeds of greenhouse plants will germinate in a temperature of 15 to 18° C. (60 to 65° F.) though seeds of warm house plants will germinate better at around 20° C. (68° F.). Some seeds germinate rapidly and seedlings may appear in 7 to 8 days, others are slow or irregular in germination, sometimes taking six months and occasionally a year or more. With the larger seeds it is usually possible to uncover one or two if germination is delayed to confirm that they have not rotted away.

Though many seeds germinate best in the dark, which is one reason for covering seed pans with paper, this is not true of all seeds as certain kinds will only germinate when they receive at least some light. There is a great dearth of accurate information on this matter, especially for the less common plants and so, when difficulty is experienced, it is wise to try some seeds in the dark and some in the light. Any shading must be removed immediately the first seedlings appear and glass coverings should be tilted slightly to admit more air, and entirely removed a few days later. Most seedlings quickly become drawn, i.e. long, thin and pale, if permitted to grow with insufficient light.

Seeds must have moisture and air for germination so the soil must be kept constantly moist but not sodden. Water is better given by semi-immersion than by overhead sprinkling.

Pricking Out

Seedlings are transplanted, or pricked out, once the seed leaves are well developed. Here, a box of seedlings is being watered overhead from a watering-can fitted with a rose

The first leaf or pair of leaves produced by seedlings are known as the seed leaves or cotyledons. In many plants they are quite different in appearance from the subsequent leaves, being unbroken in outline.

The seedlings unless sown very thinly will need to be transplanted to other trays or pans when still very small. This is known as pricking out and is best done either as soon as the seed leaves are well developed or when the first true leaves appear. As a rule the same soil or peat mixture for germination is used for pricking out, but for some strong growing plants a slightly richer compost is preferred, such as John Innes No. 1. The seedlings should be well watered the day before they are to be pricked out so that the soil around them is nicely moist. They are lifted very carefully with a sharpened stick or household fork, are equally carefully separated from one another and then replanted at least $1\frac{1}{2}$ in. apart in holes made with a pointed stick or dibber. These holes must be large enough to accommodate all the roots and the soil should be pressed firmly around the roots with dibber or fingers. When pricking out into peat compost, which is light and easily worked, many gardeners prefer to use their fingers for the whole operation.

After pricking out, seedlings must be well watered in from a can fitted with a fine rose.

Care of Seedlings

Mignonette is one of the most pleasantly fragrant of annuals. Seed is sown in spring and the seedlings either planted out in late May or grown on as pot plants in the greenhouse

After being pricked out seedlings usually need to be kept in a still, moist, fairly warm atmosphere for a few days to get established. If a propagating frame is available inside the greenhouse they can be placed in this, alternatively they can be put in the warmest part of the house and shaded from strong sunshine. The seedlings may flag a little at first but as soon as they are seen to be growing again normal treatment should be resumed. Certain plants, particularly those grown primarily for their foliage, enjoy some shade much of the time, but most flowering plants like light, especially in the early stages of their growth, and become drawn and weakly if deprived of it. One of the advantages of having shelves under the ridge in the greenhouse or above the staging is that seed trays and pans can be stood on these, so bringing the seedlings nearer the light.

Seedlings grow rapidly and need more water as they gain in size. It is usually best to use a can fitted with a moderately coarse rose which will supply plenty of water without washing the soil out. No feeding is needed.

As soon as the little plants have filled their new trays or pans they should be potted on singly into small ($2\frac{1}{2}$- or 3-in.) pots and a potting compost. Care should be taken not to break roots or damage leaves. After potting the plants should be watered and kept in still, warm conditions for a few days.

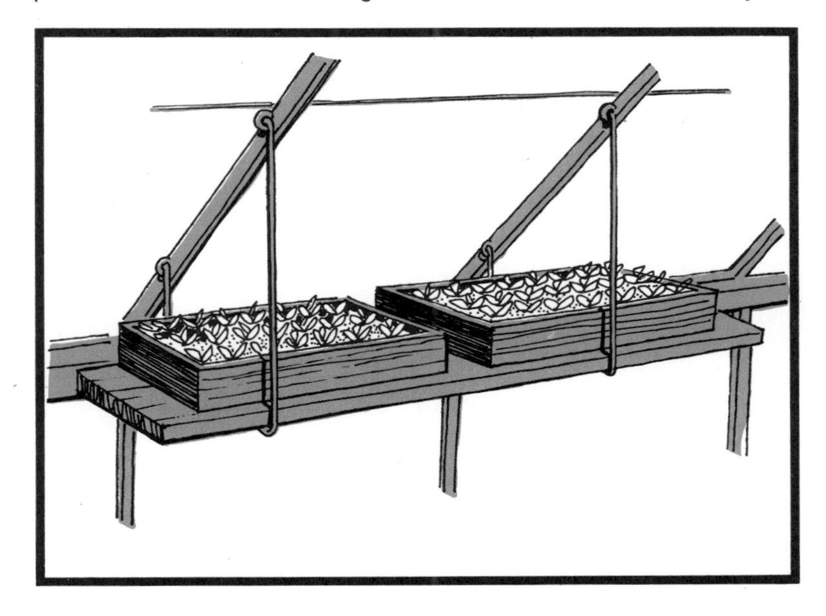

Hardening Off

If a frame is available, it will be extremely useful for hardening off plants that have been raised in a greenhouse. When the risk of frost is past they can be planted out of doors

Some plants raised or started under glass are planted out of doors later on. Care must be taken to see that the plant is properly prepared for the change in conditions, a process known as 'hardening off'.

A frame is of great assistance in hardening off, since it is possible to remove completely the protecting glass, or light, and give a degree of ventilation impossible inside a greenhouse.

First move the plant to the coolest part of the greenhouse and keep it as near the glass as practicable, giving it as much ventilation as possible without harming other plants in the greenhouse. Then remove it to a frame, continuing to increase ventilation whenever the weather is favourable. On mild days remove the light altogether by day and only replace it for the night.

It is possible, however, to harden off without using a frame. Keep the plant a little longer in the greenhouse and then put it out of doors in the most sheltered place available. Cover it at night with suitably supported brown paper or hessian and, if frost threatens, cover it by day as well.

A period of from three to four weeks is usually required to harden off a plant properly.

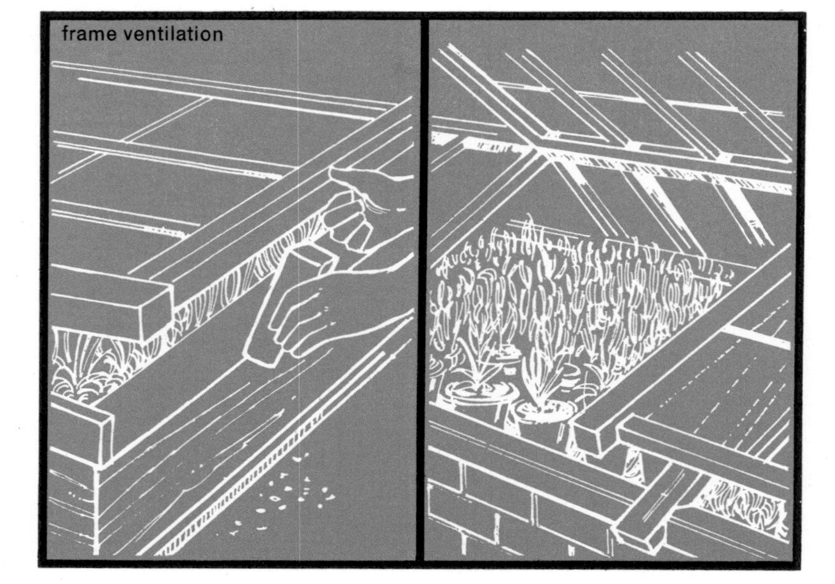

Resting and Starting

These Poinsettias are resting under the greenhouse staging during the spring. At this time of the year watering should be reduced to the barest minimum

Few plants continue to grow all through the year and some have quite long resting periods, by no means always in winter. Plants that are resting need little or no water and some actually benefit from the fullest exposure to the sun. These may be placed on a shelf near the glass in the sunniest part of the greenhouse to enjoy a good baking.

Other plants may be stored away while resting, either in their pots or, if they have fleshy roots, they may be shaken out and stored quite dry. Correct treatment during the resting period can sometimes determine the freedom with which a plant will subsequently flower.

After their resting period plants must be restarted into growth, and this, too, must be done correctly. Plants that have been shaken out of their soil are often started in damp peat in shallow trays and potted as soon as they have leaves and roots. Plants rested in their pots may simply be watered again or, if they are overcrowded, may be re-potted and then watered.

In either case water must be given rather sparingly at first and gradually increased as growth restarts, the correct temperature for the plant being maintained throughout.

Fumigation

Fumigation is an effective method of controlling pests in the greenhouse. Care must be taken when using the smoke generators and it will be necessary to keep out of the greenhouse for a few hours

In the greenhouse, pests and some diseases can most readily be controlled by fumigation. Special smoke generators can be purchased containing suitable chemicals. These, when ignited, will rapidly fill the greenhouse with smoke carrying the insecticides or fungicides with it and depositing them in a fine film on every part of the plants. Greenflies, caterpillars, whiteflies and most other troublesome insects can be controlled by this means, though for whitefly it may be necessary to repeat the fumigation at least twice at intervals of a fortnight.

To control red spider mites, which can increase rapidly under hot, dry conditions, fumigate the house with azobenzene smoke generators or use azobenzene in the form of an aerosol spray. Red spider mites are minute, rusty-coloured creatures living mainly on the underside of leaves to which they usually give a mottled, grey appearance.

A fungicide named tecnazene or TCNB can also be obtained in smoke generators and is useful for the control of botrytis or grey mould and other diseases.

Use smoke generators according to the manufacturers' instructions, first calculating in cubic feet the capacity of the greenhouse by multiplying the length, in feet, by the breadth and the height, measured mid-way between the eaves and ridge.

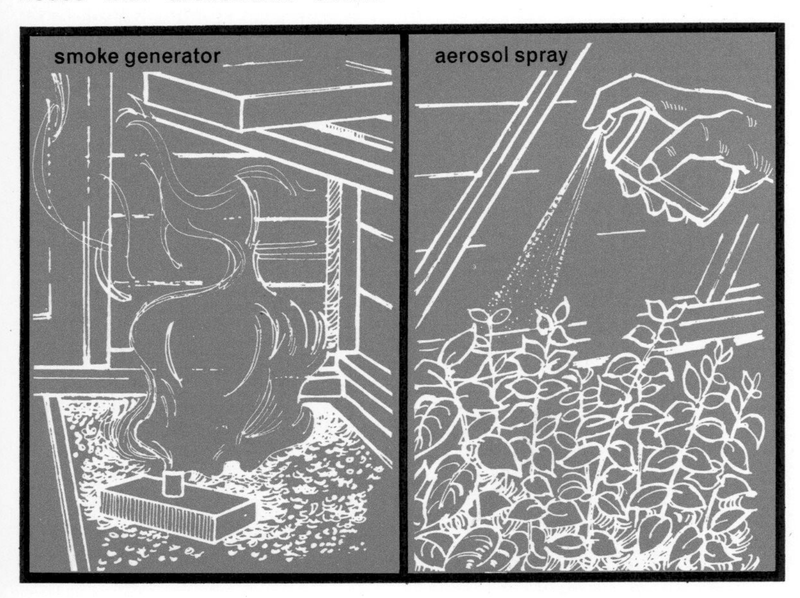

Greenhouse Hygiene

To maintain a good display of healthy plants it is important that some attention is given each year to greenhouse hygiene, which is the most effective way of controlling many plant troubles

Some pests may establish themselves in the woodwork of the greenhouse and so it is a good thing if this can be thoroughly scrubbed at least once a year using warm water, a little detergent and just a dash of any good disinfectant. At the same time glass can be washed inside and out to remove grime and green scum.

Soil or peat composts in pots and seed trays should not be used a second time since they may have become infected. One of the advantages of ring culture is that the used soil is so readily removed and the gravel aggregate can be flushed out with water.

Where crops such as carnations are grown in beds of soil these should either be renewed annually with fresh soil that has not been used previously for a similar crop, or the old soil should be partially sterilised, either chemically or by heat. The latter is best applied as steam, e.g. by suspending bags of soil over boiling water in a closed vessel, the aim being to keep the soil at 93 to 96° C. (200 to 205° F.) for 30 minutes. Formalin (40% formaldehyde) is commonly used for chemical sterilisation, 1 pint in 6 gallons of water will treat 3 bushels of soil. This treated soil should be thrown up into a heap and covered for 48 hours to trap fumes, after which it should be spread out and left until it has lost all smell of formalin. Do not use formalin in the house if it contains any plants. Proprietary sterilising chemicals are also available.

Propagating Plants

If a great many plants are to be raised from seed, cuttings or by other means, there is considerable advantage in having a greenhouse specially for this purpose or at least setting apart some section of the greenhouse for it. Young plants need to be moved frequently as they progress from seed pans to pricking out trays and then to pots, or from cutting beds to pots of increasing size. At some stages they need still, moist, warm air and at others they are better on shelves near the glass. All these varied conditions can be more easily supplied in a special house than in one which is mainly used for mature plants that have settled requirements.

The ideal propagating house will be in a particularly light position and will have a heating apparatus to meet all demands. It will be both an advantage and an economy to have one or more propagators or propagating frames installed within it as then the special conditions required by seeds and cuttings can be provided accurately and with the minimum use of fuel.

The propagating house should also be well provided with staging and shelves so that a large number of small plants can be accommodated and be tended easily. One of the commonest pitfalls of home propagation is to discover that one has inadequate room to grow on the large number of plants that can be produced from a few packets of seed.

It may be a great help to install electric lighting in the propagating house as much work is likely to be done in late winter and early spring when days are still short.

Either in or near the propagating house there must be room to carry out the necessary work. Seed trays will have to be filled with compost and sown with seeds; seedlings will need to be pricked out and later potted singly; cuttings will have to be prepared and so on. For all these jobs a solid wooden bench is ideal and it is unlikely that valuable greenhouse space will be spared for it, though some gardeners do manage with a portable bench that can be set up inside the greenhouse when required.

As a rule it is more convenient to make provision for all this ancillary work in a garden shed placed reasonably near to the greenhouse. Here potting ingredients, trays and pots can be stored, and tools kept conveniently to hand. The potting bench should stand in the window where it will receive maximum light and since much propagation is done early in the year when days are short, it will almost certainly be wise to make provision for artificial lighting and heating as well.

Propagation adds a new dimension to gardening and provides an insight into characteristics of some plants which cannot be obtained simply by cultivating mature specimens. For this reason alone it is worth carrying out some home propagation quite apart from the economy and satisfaction of producing one's own plants.

Special Methods

This summer bedding scheme uses a range of half-hardy plants which have been raised and grown on in a greenhouse for planting out when they are at their best

In addition to the methods of increasing plants by seeds, cuttings and division, there are other means suitable for greenhouse plants.

Some kinds, notably the large-leaved begonias and also saintpaulias, can be increased from leaves. With begonias, mature leaves are laid flat on the surface of a mixture of moist sand and peat. Slits are made at intervals through the main veins and these wounded veins are then held close to the compost with wire pegs like hairpins. With saintpaulias, mature leaves are detached complete with leaf stalks, and are pushed into a mixture of peat and sand until most of the leaf stalk is covered. These leaf cuttings must be kept in a warm, moist atmosphere, preferably in a propagating frame. Little plants will, in time, form from each vein incision on the begonia leaves and from the leaf stalks of the saintpaulias.

Air layering can be applied to some semi-woody plants such as rubber plants, dracaenas, and crotons. If these plants get too tall and leggy an incision is made in the bark right round the stem 2 or 3 inches below the leaves; this wound is lightly dusted with hormone rooting powder and is wrapped around with damp sphagnum moss. The moss is bound in place with fillis or raffia and is covered with thin plastic film, tightly bound top and bottom. In time roots are formed from the wound into the moss and then the whole top of the plant is severed and re-potted.

Propagating Frames

A propagating case in a greenhouse containing soft-wood cuttings of various plants including begonia, abutilon, plumbago and coleus. A sandy compost is used as the rooting medium

This is the name given to frames used mainly for raising seedlings and for rooting cuttings. Such frames are often quite small and will stand inside a greenhouse, so giving double protection and providing the very still air that seeds and cuttings enjoy. It is, of course, much cheaper to maintain the necessary temperature inside such a frame than to heat up the whole of a greenhouse to the required level.

Propagating frames are best heated from below because seeds germinate and cuttings root most rapidly when the soil is a few degrees warmer than the air. Electric soil-warming cable is excellent for this purpose. Provided the frame can be tightly closed it is not as a rule necessary to shade it except for a few seeds that germinate better in the dark. Cuttings and seedlings grow better with good light provided this does not mean that they dry out rapidly, and this they will not do in a 'close' frame, that is, one with little or no change of atmosphere.

Seeds can be sown and cuttings inserted directly in a bed of good soil, such as John Innes or peat-based seed compost spread inside the propagating frame, or they can be sown and inserted in pots, in which case it is best to plunge the pots up to their rims in moist peat half filling the frame.

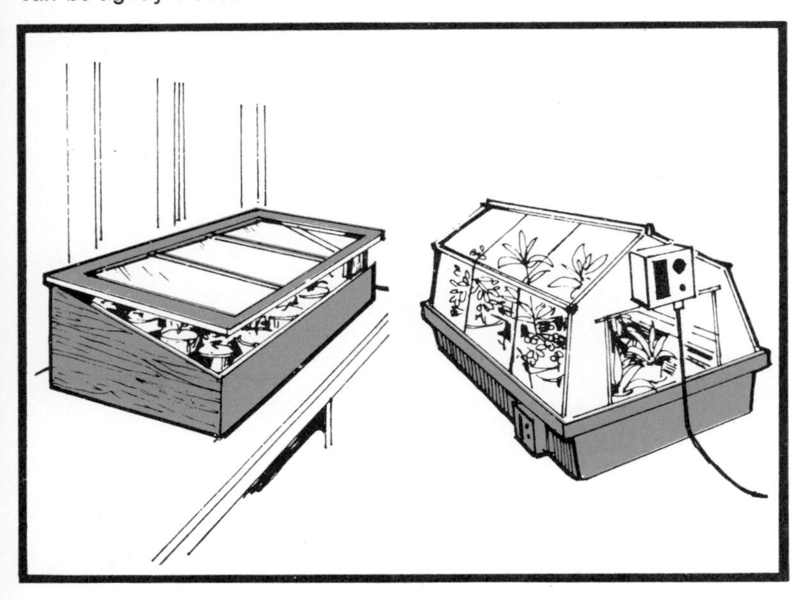

Mist Propagation

Laying the electric heating cables for a mist propagating unit. These propagators are an asset to any gardener, for they make the rooting of cuttings a relatively easy task

Mist is a device to keep cuttings fresh without confining them within a frame, box or bag, or shading them from sunshine. With free circulation of air the cuttings are less likely to succumb to disease, and full exposure to sunlight increases their rate of growth.

The moisture is provided by frequent fine sprays of water. These may be controlled by some form of time switch or by an electronic 'leaf' which is sensitive to moisture. Various forms of apparatus can be purchased and must be installed according to the manufacturer's instructions.

Mist propagators may be placed in a greenhouse or frame. Some form of soil warming is desirable. An electrical soil-warming cable may be laid 4 or 5 in. beneath the surface of the bed or the propagator may be installed over hot water pipes or other heating apparatus. A very porous compost is required, either entirely sharp sand or a mixture of sand and peat with sand predominating. This may be in a bed or in pots, pans or boxes.

If the mist is controlled by a timing device this should be set to keep the leaves constantly damp and the soil moist but not flooded.

Mist propagators can also be used to germinate seeds.

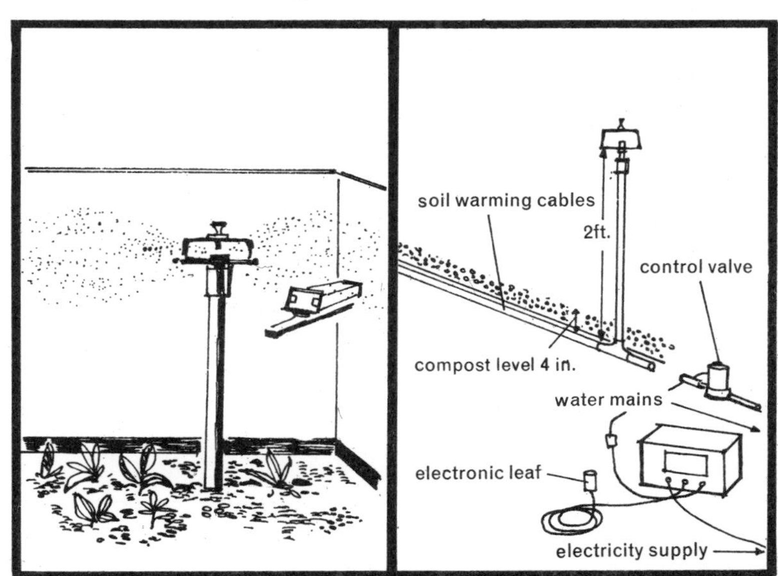

Growing in Frames

In a frame a controlled climate can be created to suit the needs of particular plants, just as it can in a greenhouse, and usually at considerably less cost. An added advantage is that when the protective lights (or other covering) are removed, plants are fully exposed to outdoor conditions in a way that is usually impossible with greenhouses. This can be of great assistance in the process of acclimatisation, called 'hardening off', which should precede the removal of plants from an artificial to a natural climate. The principal drawback of frames is their lack of headroom, which makes it impossible to grow tall plants in them or for the gardener to tend the plants with the same degree of protection for himself.

Perhaps the greatest value of frames is as an adjunct to the greenhouse. Used in this way they will serve both as an overflow for surplus plants and as a staging post for those plants that, having been reared in a greenhouse, are eventually destined for beds out of doors. In summer, with the protective lights removed, frames can provide a safe standing ground for many pot plants that are happier in the open. In winter, frames can be filled with small plants that require protection, though if this means full protection from frost then some system of heating must be provided. Considerable heating economies can be effected by covering frames with sacking or mats on cold nights. There are many winter-flowering plants which though quite hardy are often spoiled by wind and rain. In a frame they can be suitably protected without being unduly forced.

Frames are also excellent places in which to rear seedlings and root cuttings. Another use for frames is to ripen the bulbs of early-flowering plants such as tulips and sparaxis after they have finished flowering. By keeping the lights in place the rain will be kept from the bulbs. Air can be given by wedging up the lights rather than removing them or pushing them open.

When choosing and siting a greenhouse it is a good idea to choose and site the frame or frames that are to go with it, since then the whole layout can be made to look neat and efficient. It is often possible to have frames against the walls of the greenhouse where they will derive some protection from it. If the frames as well as the greenhouse are to be heated, considerable economies in installation can be made in this kind of way, perhaps by bringing heating cables or warm-water pipes from the greenhouse to the frame, or at least making certain that both can be supplied from the same source of heat. Any water supply can similarly be shared. It is desirable to have a hard path around frames and some means of stacking frame lights clear of the ground when not in use on the frames.

Types of Frame

Dutch lights are excellent for plants which require the maximum amount of light, since they are glazed with one large sheet of glass. A block of wood is used to prop up the frame light while ventilating

Frames are of many different types and sizes, but all have protective 'lights' that can be removed altogether or opened completely to allow plants inside the frame to be tended easily and to expose them to the outside atmosphere if required. Because of the comparatively small volume of air in a frame it is much cheaper to heat than a greenhouse with comparable standing space for plants, but the lack of headroom means that a frame is not suitable for tall plants.

The traditional type of frame with wooden sides and sliding lights is still one of the most popular and useful. Those used in nurseries and market gardens usually have lights measuring 6 x 4 ft., glazed with numerous panes of glass. Smaller lights, usually 4 x 3 ft., are better for gardens as they are more easily handled. Dutch lights measure 62 x 25 in. and are glazed with one large sheet of glass. They are excellent for winter crops and others requiring maximum light.

Frames may also have brick or concrete walls or be made of metal, plastic or glass fibre. Some have hinged instead of sliding lights. Some frames can be moved on metal rails and these are specially suitable for plants such as alpines which may be grown permanently in a frame, but which need only occasional protection.

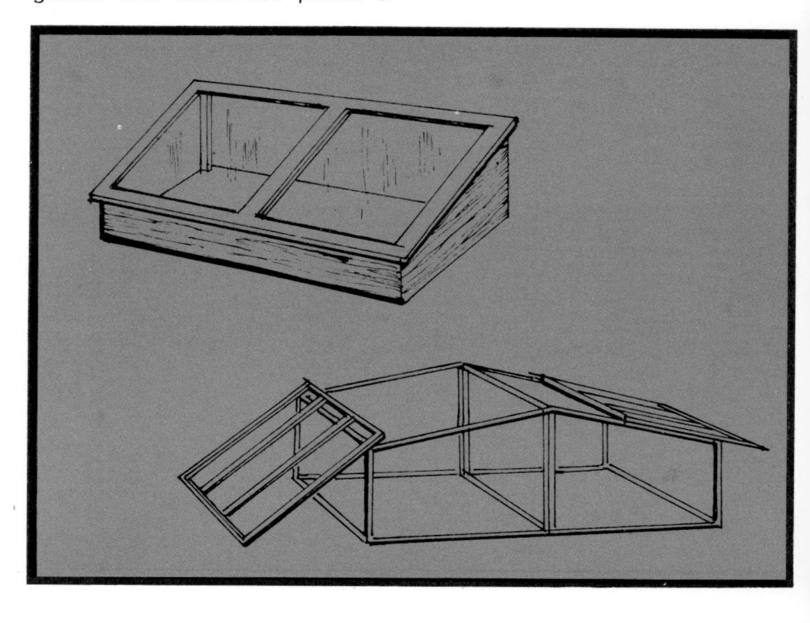

Heating, Hardening Off

When the weather is frosty it is a good idea to give some protection to frames at night by covering the lights with hessian or a similar material. This covering can be removed by day

A cold frame is one that is not heated in any way. It is most useful as an adjunct to a greenhouse, providing a place in which seedlings and cuttings raised in warmth can gradually be accustomed to the outside atmosphere. This is done by slowly increasing the ventilation until, when the weather is favourable, the lights can be removed altogether, first by day only, then later at night as well. This is known as 'hardening off'.

There are various ways of heating frames including the use of small oil heaters and hot-water pipes, but one of the most convenient methods is with special electric warming cables. These may either be buried in sand 4 to 6 in. beneath the soil's surface or they may be clipped around the sides of the frame. Soil warming is excellent for early seedlings and for rooting cuttings, air warming for plants that need frost protection in winter, and the temperature of both can be controlled by a thermostat if desired. Electric equipment in a frame must be entirely waterproof and should only be installed by a competent electrician.

Some protection can also be given by covering frames with hessian or something of the kind when the weather is frosty. This is particularly useful in spring when most frosts occur at night or early in the morning and the covering can be removed by day.

Hotbeds

A hotbed is made by placing a frame on a heap of fresh manure over which has been spread a 6-in. depth of soil and peat. This centuries-old idea can be used to advantage with many plants

When fresh manure decays it produces heat and a frame placed on top will trap the rising warmth and become a miniature hot house in which seedlings can be reared, cuttings rooted and warmth-loving plants given an early start.

Fresh strawy horse manure is the best material with which to make a hotbed. Failing this, well-wetted straw can be used treated with one of the advertised preparations for converting straw into mushroom compost. Another possibility is a mixture of autumn leaves, lawn mowings and weeds (especially nettle tops which rot well). Whatever is used should be built into a conical heap and left for about a week when it must be turned, well mixed and any dry parts watered. It will soon start to ferment vigorously and the temperature may rise considerably. The heap should then be turned again. A few days later it should be built into a flat-topped heap at least 18 in. longer and wider than the dimensions of the frame that is going to stand on it, and 2 ft. deep after having been well trodden down. A 6-in. depth of good soil or peat is spread on top and the frame and frame light placed in position. When the temperature well down in the heap has dropped to about 25° C. (77° F.) seeds can be sown, cuttings inserted or plants introduced. The warmth of the heap will continue to drop, and should remain effective for about 8 weeks.

Plants for Unheated Greenhouses

Only in the very mildest areas can an unheated greenhouse be expected to exclude frost throughout the winter. At this season, therefore, such a house is only suitable for plants that will withstand some degree of freezing. From about April to October it is very different, for then frosts are usually of short duration, sun heat can raise the temperature of the greenhouse to quite high levels by day even when it is cold outside, and stored heat will usually carry the house through the nights with little risk of the temperature dropping below freezing point. For these reasons during this period an unheated greenhouse can be used to grow quite a wide range of tender plants and also to raise seedlings, root cuttings and propagate plants in other ways.

If a conservatory is placed against a dwelling house, with perhaps french windows or a door leading directly into one of the rooms, it may derive quite a lot of warmth from the house, though even so it is unlikely to be fully protected against frost in winter unless it has some heating of its own. In winter such a conservatory can be used for the display of many hardy plants (including bulbs) in flower, also for nearly hardy plants such as cinerarias and winter-flowering primulas and for hardy or nearly hardy foliage plants such as clipped bay trees, fatsias, ivies, *Dracaena australis*, and the hardy palm, *Chamaerops humilis*. In summer there are few greenhouse plants that cannot be accommodated temporarily in it.

There is, of course, a half-way step from the completely unheated to the heated greenhouse, and that is the house with strictly temporary heating to be used only in emergencies. This is probably the way most beginners approach the problem, making do with portable oil heaters or something of the kind simply to exclude the worst frosts or to enable a start to be made with seed sowing a little earlier than would be wise in a completely unheated house. It is a very sensible compromise provided its limitations are realised, and it is one that, as experience and enthusiasm grow, is likely to lead eventually to a more generous level of heating.

Whether completely unheated or not, greenhouses which are to be run 'cold' most of the time must be sited with special care. They should be placed to catch all the sunshine that is going, but if they can also be sheltered from north and east by a building or hedge so much the better. The microclimates created in such places can be several degrees warmer than in a fully exposed position a few yards away.

Care during watering can also save the loss of a few degrees as water when evaporating takes heat from its surroundings. Therefore capillary benches should be allowed to dry out in the autumn and the plants grown with a minimum of water, care being taken not to spill water even on the path. Water during the morning to give any heat loss a chance to be made good before temperatures drop in the early evening.

Annuals

By making a succession of sowings, it is possible to have schizanthus in flower throughout the spring, summer and autumn. This lovely free-flowering annual is easy to grow

Any hardy annual can be grown in an unheated greenhouse and some kinds, such as amaranthus, dimorphotheca, mignonette and schizanthus, make excellent pot plants. Seed of all these can be safely sown in late March or early April to give flowers in summer.

Even half-hardy annuals can be grown in an unheated greenhouse if sowing is delayed until April, or if seedlings are raised in a heated propagator or are purchased in April or May and potted. Among the best kinds for pot culture are antirrhinum, arctotis, *Begonia semperflorens*, Chabaud carnations, annual dianthus, diascia, felicia, heliotrope (strictly a perennial but easily grown as an annual), lobelia (especially the trailing varieties), molucella, nemesia, petunia (especially the large flowered and double varieties), salpiglossis, Ten Week stocks, ursinia, verbena and zinnia. *Cobaea scandens* and ipomoea varieties Flying Saucers or Heavenly Blue can be grown in the same way as climbers and will rapidly run up canes or spread along wires.

All these annuals are sun lovers and require no shading. They should be germinated and pricked out in J.I.S. or peat seed compost and then potted singly in J.I.P. No. 1 or peat potting compost. All should be watered fairly freely throughout and may be fed once a fortnight with weak liquid manure from the time they form their flower buds. Remove faded flowers.

Biennials

The biennial stocks, such as the East Lothian and Brompton types, are more likely to survive the winter and will flower more freely when they are raised in an unheated house

Some biennials make excellent pot plants for an unheated greenhouse or conservatory. Wallflowers can be had in bloom several weeks earlier than if grown out of doors. East Lothian and Brompton stocks are far more likely to survive and flower freely than they would in any but the mildest places in the open. The dwarfer forms of the Canterbury Bell make first-rate pot plants and its relative the tall Chimney Bellflower, *Campanula pyramidalis*, will produce striking 5-ft. spikes of bloom provided it does not get frozen too severely. Sweet Williams are completely hardy and can make a brilliant display.

Seed of all these can, if wished, be germinated outdoors in May though it is usually safer to sow in a frame. Seedlings can also be grown on for a while outdoors or planted out in a frame, but should be potted individually in J.I.P. No. 2 or peat potting compost before they get too large. Thereafter they can stand out of doors or in a frame until October, after which they should be rehoused and water should be applied much more sparingly. The plants are discarded after flowering. Some of the winter-flowering primulas, and particularly *Primula malacoides* and *P. kewensis*, can be grown in a similar manner except that seed should be germinated in pots and the seedlings grown on in pots throughout. They need only frame protection in summer but should go into the greenhouse by late September.

Herbaceous Plants

Some of the herbaceous plants, such as the astilbe shown above, make excellent pot plants for an unheated greenhouse but an annual re-potting is usually essential

A number of herbaceous plants grow well in pots and even in an unheated greenhouse can be in bloom several weeks ahead of their normal flowering time. Plants can either be purchased or lifted from the open ground and transferred to pots. This is usually done in October or November, though it can also be done in spring, in which case it is unwise to bring the plants into the greenhouse before the following autumn. Throughout the summer the plants can stand outdoors.

Since most herbaceous plants grow rapidly annual re-potting is usually essential. As the plants get too large for comfort they can be split up into smaller pieces. Water sparingly in winter, fairly frequently in spring and summer and feed with weak liquid manure about once a fortnight from June to August inclusive. Watering can be reduced by plunging pots to their rims in soil, sand or peat when they are put out of the greenhouse after flowering.

Among the most useful kinds are primroses, polyanthus and doronicum for spring, dicentra and astilbe for early summer. Francoa, though not completely hardy outdoors, usually survives in an unheated greenhouse and so does the Mother of Thousands, *Saxifraga tomentosa*, an attractive trailing plant. Violas and pansies can be grown in quite small pots but are often difficult to over-winter. Both are readily raised from seed.

Bulbs in Pots

For spring flowering, this novel pot was planted with tulip bulbs in early autumn. The pot requires careful planting as each bulb must be placed with its 'nose' by a hole

Many different types of bulbs can be grown successfully in an unheated greenhouse. For spring flowering only hardy kinds should be chosen, such as narcissus, tulip, hyacinth, crocus and the early bulbous-rooted irises, since they will have to be potted in early autumn and will make their growth in winter. For summer flowering some more tender kinds can be grown, such as tigridias, tritonias and zephyranthes. These need not be potted until March or April, when even an unheated house should provide frost protection.

In most cases the bulbs (or corms) can be planted almost shoulder to shoulder, but a little more space should be left between the very small irises, tigridias, tritonias and zephyranthes. For all kinds either J.I.P. No. 2 or peat-based compost is suitable.

All should be watered rather sparingly at first, then with increasing freedom as growth proceeds until the leaves begin to turn yellow and wither away, which they will in May and June for spring-flowering bulbs and in early autumn for most summer-flowering bulbs. This is the signal to reduce watering until the soil is quite dry, when the bulbs can be removed from the soil and stored until re-potting.

It is an advantage if narcissus, tulips and hyacinths can be plunged under 2 inches of sand or peat in a shady place outdoors for 8 to 10 weeks before they are brought into the greenhouse.

Camellias

There are many lovely varieties of *Camellia reticulata*. All are vigorous plants and ideal for training on wires or trellis. Those with double flowers can be grown to greater perfection indoors

All kinds of camellia make excellent pot or tub plants for unheated greenhouses and the very vigorous varieties of *Camellia reticulata* can be trained on the back wall of a lean-to greenhouse or conservatory. Varieties with large double flowers and those with double flowers of very formal shape can be grown to greater perfection in a greenhouse or conservatory than is usually possible out of doors where these flowers are often damaged by frost, wind or heavy rain. For the same reason this is an excellent way to grow the beautiful varieties of *C. sasanqua* which flower in autumn and winter.

It is best to start with young plants in pots, which can, if wished, be purchased in flower. If any are to be planted permanently in the greenhouse a bed of lime-free loam mixed with plenty of peat should be prepared for them. Pot plants are grown in a mixture of 4 parts by bulk lime-free loam, 2 parts peat and 1 part coarse sand with 4 oz. John Innes base fertiliser added to every bushel.

Stand pot or tub grown plants outdoors from May to October in a sheltered place. Ventilate permanently planted camellias very freely in summer and shade from direct sunshine. Water freely in spring and summer, moderately in autumn and winter. Feed every 10 to 14 days from May to August with weak liquid manure. If plants grow too large stems can be thinned or shortened in late May.

Hydrangeas

These popular pot plants are best renewed annually from cuttings. No chalk or limestone should be added to the compost if blue flowers are required and a 'blueing' compound may be used

These make excellent pot plants, for which they are best renewed from cuttings annually, though old plants can be grown in large pots or tubs in a border of soil within the greenhouse.

Prepare cuttings in March or April from firm, non-flowering stems. Insert the cuttings in sand and peat in a greenhouse or frame with a temperature of 10 to 15° C. (50 to 60° F.). When well rooted pot singly in 3-in. pots in J.I.P. No. 1 or peat-based potting compost. If blue flowers are required, make sure there is no chalk or limestone in the compost, and use a special hydrangea 'blueing' compound according to instructions.

Water freely and grow on in a greenhouse, with shade from strong sunshine. No artificial heat is required from May to October and plants can be put in a frame from June to September.

Pot on into 5-in. pots and J.I.P. No. 2 or peat potting compost when the smaller pots are full of roots and remove the tip of each plant about a fortnight later to encourage branching. Return them to the greenhouse in October and at the same time pot on the largest plants to 6- or 7-in. pots in similar compost.

Water rather sparingly from November until January and maintain a minimum temperature of 7° C. (45° F.). Then raise the temperature to 10 to 13° C. (50 to 55° F.) and water more freely for early flowers.

Other Shrubs

The unheated greenhouse is useful for providing winter protection for some of the early-flowering and slightly tender shrubs. *Rhododendron edgworthii*, shown above, is just such a plant

Cytisus

Many other shrubs grow well in pots in an unheated greenhouse. Some, such as *Cytisus canariensis*, the shrub with scented yellow flowers in spring commonly known as 'genista', are a little tender but are usually safe in an unheated greenhouse provided they are not overwatered in winter. So are *Deutzia gracilis*, with sprays of small white flowers in spring and *Prunus triloba flore pleno*, a pretty relation of the almond with double pink flowers in spring. The Bridal Wreath, *Spiraea thunbergii*, is frequently grown as a white spring-flowering pot plant, and *Choisya ternata*, the Mexican Orange Blossom, is also very suitable.

Rhododendrons and azaleas make compact root balls and even quite big specimens can be grown in large pots or small tubs. Some rhododendrons are a little tender or flower very early, e.g., Christmas Cheer, *R. moupinense*, *R. ciliatum*, *R. praecox*, and like protection.

All shrubs should be potted in November, usually in J.I.P. No. 2, though for rhododendrons and azaleas no chalk or lime should be included. They are watered moderately in winter, freely in spring and summer and may stand outdoors from May to October. Feed fortnightly from June to August with weak liquid manure and re-pot annually in November. Genista, spiraea, deutzia, prunus and choisya can be pruned after flowering. Rhododendrons and azaleas should only have the faded flowers removed.

The Alpine House

An alpine house is a greenhouse devoted solely or mainly to plants which might otherwise be grown in a rock garden. Many of these will be genuine mountain plants but not necessarily all of them. Some will come from rocky places at low altitudes, others may simply be small plants which look right in the company of true rock plants. There are two major reasons for growing such plants in an alpine house: one, that it enables many of them to be grown to greater perfection than is possible in the open garden; the other that as many flower very early in the year it is easier to enjoy their beauty in the comfort of a greenhouse than it is out of doors. Since many of these plants come from very cold places it might be supposed that greenhouse protection is the last thing they would require. In fact, in their native habitats they are often covered deep in snow for the greater part of the year, emerging when the snow melts for a few weeks of hectic life and then returning to hibernation again. In lowland gardens they miss their snow covering, never become properly dormant, start to grow much too early and then have their flowers and possibly their leaves damaged by frost, wind and heavy rain. All these hazards can be overcome in a well managed alpine house.

A frame or series of frames is almost a necessity as an overflow for an alpine house, a place in which to keep plants while they are dormant or making little contribution to beauty. Frames have the advantage over greenhouses that the protective glass can be completely removed if desired, which can be very useful in summer to keep the temperature down. Even with full ventilation a greenhouse is usually a good many degrees warmer than the outside air on sunny days. An alpine house must be well provided with ventilators, all along the ridge if possible, and at the sides as well, so that when required a current of air can flow right through it. Staging is almost essential to bring the plants closer to eye level where they can be appreciated more fully and cared for most easily. It is an advantage if this staging has a solid, not a slatted, top on which a layer of pea gravel can be spread. This will absorb moisture and on sunny days will help to maintain the cool moist atmosphere which most rock plants enjoy.

Though most alpine houses are unheated this is not really ideal. Artificial warmth may not be required very often but there are occasions when it is desirable to exclude frost, certainly very severe and prolonged frost, and others when the air gets too moist and a little drying out is advantageous. The heating is unlikely to be required for more than that, so can be on a much more modest scale than in most greenhouses and will be correspondingly economical to run.

Alpine House Management

An unheated greenhouse may be used exclusively for the cultivation of spring-flowering alpines or rock plants and it is then known as an alpine house

All alpines enjoy a good circulation of air, so ventilation of the alpine house must be as free as circumstances permit. From April to October ventilators will be wide open most of the time and frame lights will only be required to keep off excessive rain, particularly from bulbs and tubers that are at rest. Even in winter it will be possible to keep the top ventilators open much of the time, only closing them when severe cold threatens to lower the temperature well below freezing point. In autumn especially, when the air is often excessively wet, it may be necessary to use a little artificial heat while keeping the top ventilators open to maintain a steady circulation of air and to dry it out a little. This is particularly necessary for plants with leaves densely covered in hairs or down which can rot away from excessive moisture.

If frames are available it may be possible to use the alpine house itself mainly for display, moving many plants into it just before they are about to come into flower, or when their foliage is at its best. In a frame, pans can be plunged to their rims in sand, pea gravel or peat and this will help to keep them moist and so reduce the frequency of watering.

Although most plants will appreciate all the light possible from October to April, from May to September most will benefit from light shading, either with muslin or green shading compound.

Small Bulbs

The delicate flowers of many of the early-flowering bulbs can be grown to perfection in an alpine house where they are protected from wind and rain. *Iris reticulata* Harmony is a delightful choice

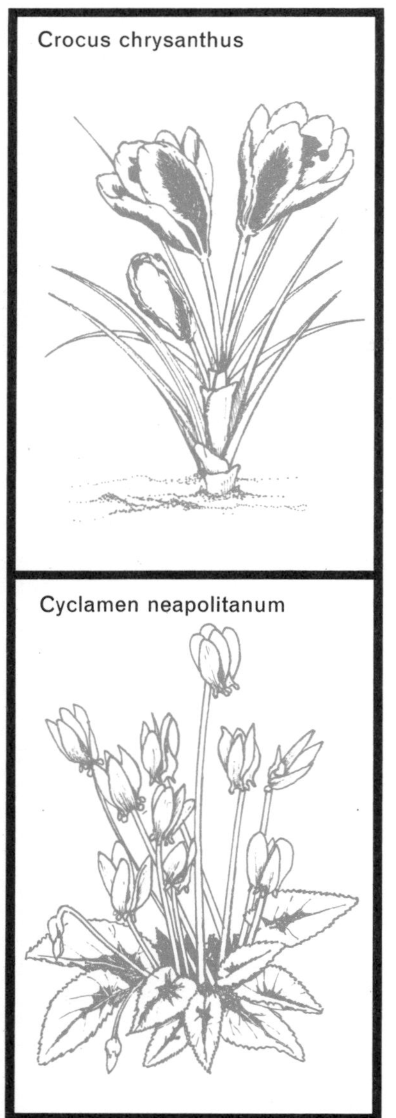

It is very early flowering bulbs that best repay cultivation in an alpine house since here their fragile blooms are protected from wind and rain. This applies to such irises as *Iris histrioides*, blue, *I. reticulata*, violet-purple and *I. danfordiae*, yellow; also to many crocuses, especially the numerous varieties of *Crocus chrysanthus* in orange, yellow, blue, purple and white. The large-flowered snowdrops and the early-flowering snowflake, *Leucojum vernum*, can be grown in pots, and so can the varieties of *Anemone blanda*, which may be blue, pink or carmine. Others worth their space are small species tulips such as *Tulipa violacea*, *T. orphanidea* and *T. batalinii*, the small fritillaries and *Tecophilaea cyanocrocus*.

The small-flowered cyclamen species, such as *Cyclamen orbiculatum*, which flowers in winter and *C. neapolitanum*, which flowers in autumn, make good alpine house or frame plants, and if given sufficient room in their pans they will increase themselves by self-sown seeds.

All these like a mixture of 2 parts peat, 1 part loam and 1 part sand. All should be removed to a frame after flowering or can be plunged to their rims in any sheltered place. There they will die down completely. The bulbs can be shaken out and re-potted in August, but the cyclamen are best left undisturbed, being transferred to wider pans as necessary. Seedlings can be prized out gently and re-potted separately.

Campanula and Allies

Although not easy to grow out of doors, *Edraianthus pumilio* makes an excellent pot plant for the alpine house where it can be protected from winter dampness

Many of the small species of campanula make good plants for the alpine house. These include *Campanula allionii* with quite large, purple bell-shaped flowers in June; *C. elatines*, a creeping plant with downy leaves and starry, blue-purple flowers in July; *C. waldsteiniana* only 3 in. high with starry violet-blue flowers in summer. *C. warleyensis*, a dainty little garden hybrid with small double flowers on thread-like stems in summer, and *C. zoysii* with pale blue flowers, tubular but crimped at the mouth, produced in June.

Related to these are the various species of edraianthus and phyteuma. The former have bell-shaped purple flowers which in *Edraianthus pumilio* and *E. serpyllifolius* are produced in May and June on slender stems.

Phyteumas are known as horned rampions and have unusual flowers rather like clusters of long necked onions. One of the most striking is *Phyteuma comosum*, which makes a little hummock of prickly leaves with lilac and purple flowers in June.

Most of these will grow in a mixture of 2 parts peat, 1 part loam and 1 part sharp sand. Water them fairly freely in spring and summer, sparingly in winter. All, except *Campanula warleyensis* which must be divided in spring, can be increased by seed. Campanula seeds need light for germination.

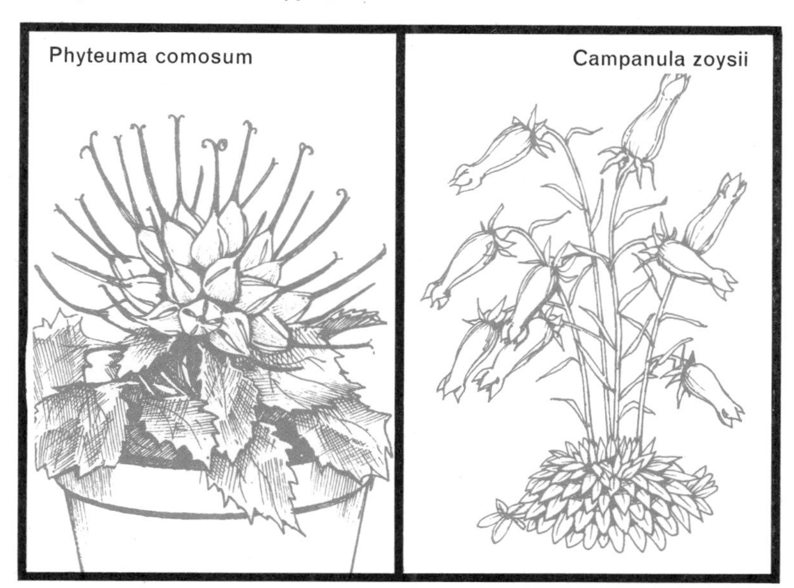

Cassiope and Relatives

Andromeda polifolia produces its clusters of bell-like flowers in May. This charming dwarf shrub grows to about 12 in. and should be placed out of doors in summer if possible

The cassiopes are small evergreen shrubs related to the heathers but considerably more difficult to grow. *Cassiope lycopodioides* and its varieties and hybrids are the most satisfactory and beautiful, capable of completely covering a pan with a close hummock of small evergreen leaves and bearing white urn-shaped flowers in April.

Nearly related to these and forming similar mats or tussocks covered in small evergreen leaves are the phyllodoces. *Phyllodoce empetriformis* has red and purple flowers and *P. nipponica* white flowers and both these excellent kinds flower in April and May.

Then there is *Andromeda polifolia*, another of the heather tribe, a little evergreen bush with clusters of pink urn-shaped flowers in May. It tends to be a little straggly in habit but there are improved forms which are much more compact and are to be preferred for pot or pan cultivation.

All these little shrubs should be grown in a mixture of 4 parts peat and 1 part each of lime-free loam and coarse sand. They should be watered fairly freely in spring and summer, sparingly in autumn and winter. In summer they are happiest out of doors in a cool sheltered place protected from direct sunshine and frequently syringed to keep their leaves moist. All can be increased by seed, or by cuttings in July and August rooted in sand and peat.

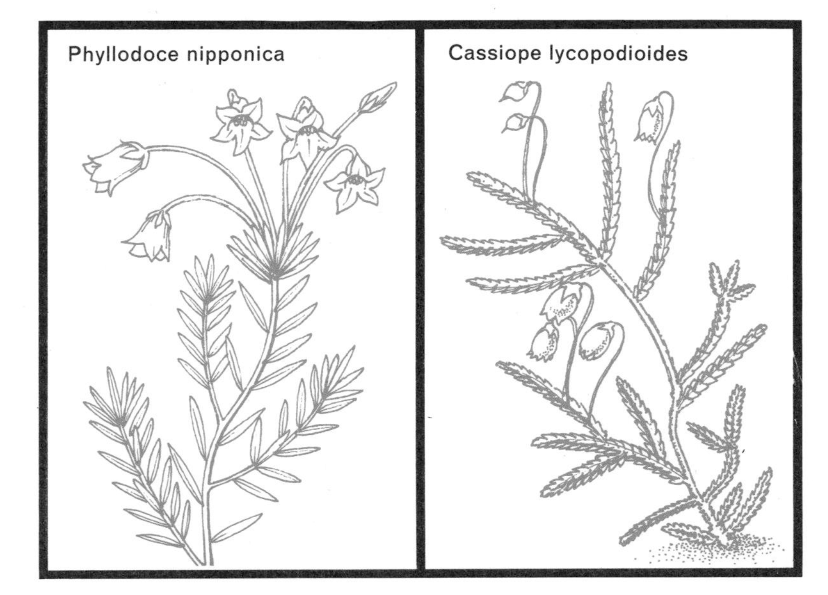

Gentians

The large trumpets of *Gentiana sino-ornata* are produced in September. All the gentians should be grown in a lime-free compost and most resent any root disturbance

There are gentians to flower in spring, summer and autumn, and all the prostrate or tufted kinds make excellent pot plants. A favourite is *Gentiana acaulis*, with large, narrowly trumpet-shaped, deep blue flowers on tufts of evergreen leaves in April or May. *G. sino-ornata* and *G. macauleyi* produce flowers of similar size and colour in September but have trailing stems which die down in winter. *G. septemfida* is a sprawling plant bearing clusters of deep blue flowers in late summer. *G. verna* has quite small flowers in April and May but the blue colour is particularly intense. There are many other species and hybrids. Most gentians dislike lime, though there is a considerable difference between species and both *G. acaulis* and *G. septemfida* may sometimes be found thriving in quite chalky soil. But for the alpine house it is best to grow them all in lime-free composts and to use lime-free water. A mixture of 2 parts peat and 1 part each of lime-free loam and sand will suit most, but for the autumn-flowering kinds the proportion of peat can be doubled.

All need to be watered freely in spring and summer and kept moist in autumn and winter. Species can be increased by seed, but hybrids and garden varieties must be increased by careful division in March, or immediately after flowering for the spring-blooming kinds. Since most kinds resent root disturbance division should only be done when essential.

Lewisias

Waxy, delicately shaped flowers in May and June are the rewards to any gardener who introduces the lewisias to his alpine house. *Lewisia howellii* is illustrated above

In addition to the wild lewisias, such as *Lewisia howellii* with sprays of salmon flowers in May and *L. tweedyi* with large apricot flowers in June, there are a number of garden hybrids which seem easier to grow. These show considerable variation in colour and flower size but most flower freely and make excellent pot plants for an alpine house. They all need specially well-drained pots and like a mixture of equal parts lime-free loam, peat and coarse sand. Most should be watered sparingly in autumn and winter, fairly freely in spring and summer, but *Lewisia rediviva* dies down completely in July and from then until growth starts in December should be kept quite dry. In summer all appreciate light shading. All can be increased by seed though seedlings of the hybrids may vary in habit and colour. Some can also be divided in spring, but others make a single rosette of leaves on top of a carrot-like root which cannot be divided. Sometimes leaf cuttings can be induced to root in summer.

Another rosette forming plant which makes a good pot specimen is *Morisia monantha*. The handsome rosettes of dark green evergreen leaves are studded in May with bright yellow flowers. This is an easy plant to grow in a mixture of equal parts loam, peat and sand. It should be watered fairly freely in spring and summer, very sparingly in autumn and winter and can be increased by root cuttings in June.

Hardy Orchids

Pleiones are hardy orchids and will do equally well in a cold frame or a warm greenhouse. They make good window-sill plants provided they receive sufficient sunheat

Several hardy orchids can be grown successfully in the alpine house. Among the most beautiful of these are the pleiones with pink or white flowers rather like small cattleyas in shape and freely produced in spring. They can be grown in well-drained pans in a mixture of 3 parts lime-free loam, 2 parts peat and 1 part sharp sand with a sprinkling of bonemeal. Pleiones are grown from bulb-like structures (pseudobulbs) which should be buried to half their depth in this compost, and 3 or 4 in. apart. Water sparingly at first then freely as growth progresses, but keep almost dry and in a frost-proof place in autumn and winter when they are at rest. Increase is by division of the pseudobulbs when re-potting after flowering.

Bletilla striata, sometimes known as *Bletia hyacinthina*, has small magenta flowers similar to those of pleione in form, but produced in loose 9-in. spikes in July and August. It should be grown in a mixture of 2 parts peat and 1 part each lime-free loam and sand, needs to be well watered in spring and summer, and shaded from direct sunshine. Like pleione it should be kept nearly dry and frost free in autumn and winter.

Other orchids that may be tried are *Cypripedium calceolus*, the chocolate and yellow Lady's Slipper Orchid, and *C. reginae*, the rose and white Moccasin Flower, both flowering in early summer. Soil and treatment are as for bletilla, but the plants are slightly hardier.

Oxalis and Others

Produced on short stems set amidst dainty leaves, the large pink flowers of *Oxalis adenophylla* brighten the months of May and June. Oxalis appreciates some shade while in bloom

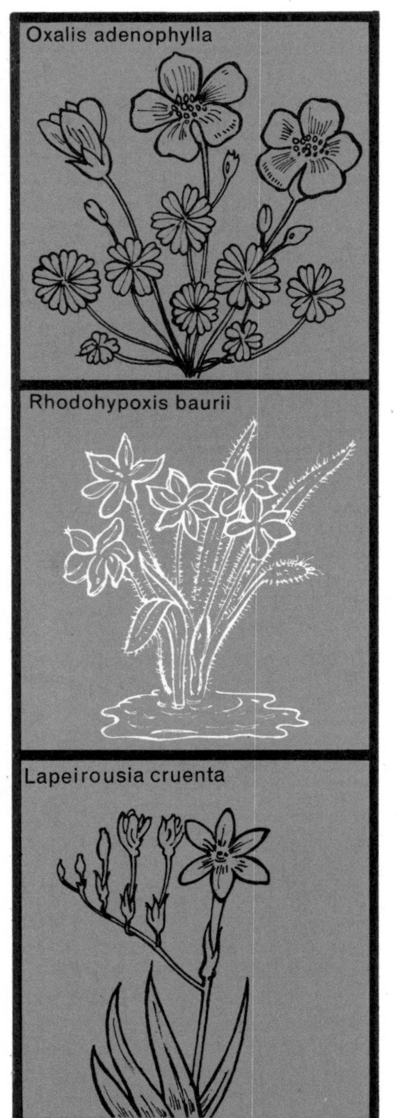

Oxalis adenophylla and *O. enneaphylla* are similar tuberous-rooted plants with deeply-divided grey-green leaves and large flowers, lilac-pink and white respectively, on short stems in May and June. Both will grow well in a mixture of 2 parts peat, 1 part each loam and sand. They should be watered freely from April to August, but as the leaves yellow and die down water must be reduced and the soil kept just moist, no more, in autumn and winter. Though they are sun lovers they appreciate a little shade while in bloom.

Similar treatment suits *Rhodohypoxis baurii* except that the loam must be lime free and no water at all is required from October to March. This delightful little plant has bulbous roots, grassy leaves and carmine flowers like fire-flies on 2-in. stems from May to September.

Lapeirousia cruenta must also be kept quite dry in late autumn and winter. It produces its 9-in. sprays of small, starry, orange and crimson flowers in August and September and will grow in a mixture of equal parts loam, peat and sharp sand.

It is grown from corms which should be potted in March, five to seven in each 5-in. pan. Water sparingly at first, freely when growth appears, but discontinue watering when the leaves die down. It is wise to keep the pans in a frost-proof greenhouse or frame in winter as the little corms are none too hardy.

Primulas

Primula pubescens Mrs J. H. Wilson breaks into colourful heads of flowers in May. Increase this attractive plant by dividing the roots in July, or by sowing seed when ripe

The primulas most suitable for cultivation in pots or pans are the European species related to the auricula. These include *Primula auricula* itself, a plant with leathery leaves more or less covered in white meal for which reason it is sometimes called Dusty Miller. The primrose-like flowers are borne in clusters on 6-in. stems in April or May and in many colours.

Primula pubescens is similar but smaller and neater, usually violet, lilac, crimson or white. *P. viscosa* is closely allied to this and also has a range of colours. *P. allionii* produces its rose-pink flowers in clusters on very short stems. It has a pure white variety. *P. glutinosa* has violet-blue flowers on 3-in. stems and *P. marginata*, with lavender flowers, is twice that height and has serrated silver-edged leaves. There are several varieties differing in depth of colour. Others in the same group are *P. clusiana* and *P. villosa*, both with carmine, white-eyed flowers.

All these flower in spring and can with advantage be outside in a fairly cool, partially shady place from June to October. All will grow in a mixture of equal parts loam, peat and coarse sand with plenty of broken crocks or small pebbles for drainage in the bottom of each pot or pan. They should be watered freely while in growth and never allowed to get really dry. They can be grown from seed or be increased by division as soon as they have finished flowering.

Saxifrages

Saxifraga grisebachii wisleyensis, one of the aptly named Silver Saxifrages, is an outstanding and highly decorative plant for the alpine house

All the early-flowering Cushion Saxifrages make beautiful specimens for the alpine house and as they spread they can be moved from pots to pans of ever greater width to display their full beauty. This applies to white-flowered *Saxifraga burseriana* in all its varieties, and to the scores of garden raised hybrids in white and various shades of yellow and pink. The Silver Saxifrages, such as *Saxifraga aizoon, S. cochlearis, S. cotyledon, S. longifolia*, and hybrid varieties like Tumbling Waters, are also excellent and flower in May and June. Then there is *S. grisebachii* with hummocked rosettes of silvered leaves from which in March arise crozier-like flower stems set with crimson bracts.

All these and many other kinds will thrive in the same soil mixture as that recommended for primulas but with a little powdered chalk added for the Silver Saxifrages. All like some shade in summer. The pans can be moved to a cooler place as the weather gets hot.

All should be watered freely in spring and summer, but the drainage must be good. In winter they should be kept rather dry.

The garden varieties are increased by careful division after flowering. Rosette-forming types can also be increased by detaching rosettes in June and inserting them as cuttings in sand and peat. Species are increased by seed, which is the only way of propagating *S. longifolia*.

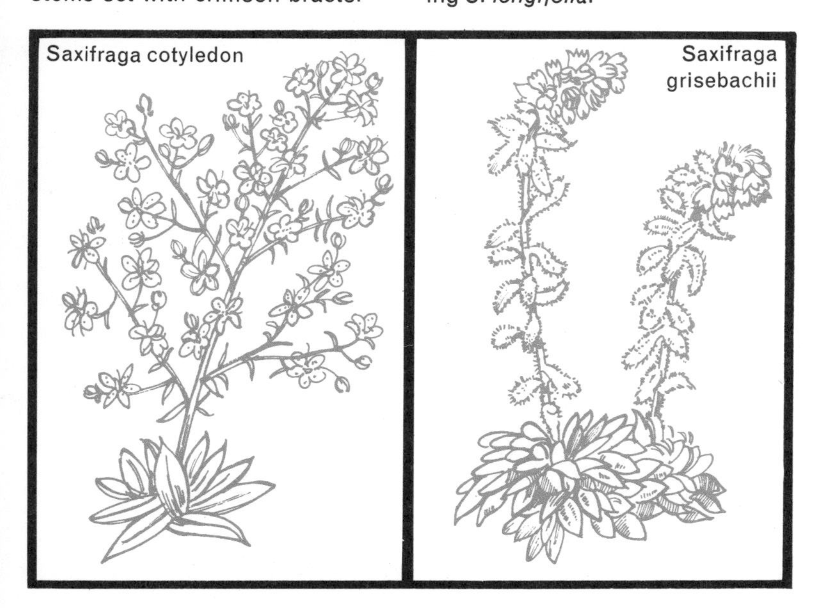

Saxifraga cotyledon

Saxifraga grisebachii

Shortia and Others

Fringed petal edges to trumpet-shaped flowers make *Schizocodon soldanelloides* a must for alpine-house enthusiasts. These unusual flowers appear in March or April

Shortia uniflora

Soldanella alpina

Schizocodon soldanelloides

The shortias are woodland plants, happier in a shady frame or plunge bed in summer than in an alpine house, but benefiting by being brought into the alpine house for a few weeks while flowering. They make low clumps of rounded shining evergreen leaves and have broadly bell-shaped white or pale pink flowers with fringed petals on 6-in. stems in April. *Shortia galacifolia* and *S. uniflora* are the two kinds grown. They need a mixture of 4 parts peat to 1 each of lime-free loam and coarse sand, and should be watered freely with lime-free water from March to September. Syringe daily with water in summer and only re-pot every second or third year, after flowering.

Soldanellas also have bell-shaped flowers with fringed petals, but they are smaller and the rounded leaves die down in winter. *Soldanella alpina* and *S. montana* both have lavender-blue flowers in April. Grow them in a compost of 2 parts peat, 1 part loam and 1 part coarse sand and treat as shortias. Slugs attack them readily and can be controlled with slug killer. *Schizocodon soldanelloides* looks like a pink and white soldanella, is related to shortia and requires exactly the same treatment.

Ramonda and haberlea make flat rosettes of leathery leaves from which spring in May and June short stems carrying clusters of lavender-blue or white flowers. They can be grown in the compost recommended for soldanella.

Plants for Heated Greenhouses

As soon as some form of heating is introduced the greenhouse owner's control over climate is increased and with it the range of plants which can be grown. Nevertheless, it is too simple a view to regard plants as divided into two groups, those that are hardy and will withstand frost, and those that are tender and are injured or killed by frost. Many tropical plants will die long before temperatures anywhere near freezing point are reached and the so-called hardy plants also differ greatly in the amount of cold they will stand. Even plants closely related and very similar in appearance may differ in the temperature ranges they will survive and only experience can prove just what those temperatures are. When in doubt it is wise to err on the side of keeping plants a little too warm than letting them become too cold.

In the following pages temperatures are referred to in four classifications, those of the cold greenhouse, cool greenhouse, intermediate greenhouse and warm greenhouse. These terms are explained on page 32, and though it is not necessary to adhere rigidly to any of these regimes they do give an accurate indication of the temperature ranges which particular plants prefer. There is no truth in the idea that warm-house plants are any more difficult to grow than those requiring lower temperatures.

Under glass plants often react more violently to direct sunshine than they would in the open. For this reason many plants regarded as sun lovers will not object to, and may actually appreciate, light shading from direct sunshine from about May to September. This can be convenient because although it is possible to shade plants individually it is much easier to spray the glass on the sunny side with a shading compound or to pin muslin under the rafters and leave it there for several months.

The provision of a correct degree of humidity is much more difficult. Most plants thrive in the kind of atmosphere gardeners usually describe as buoyant, which really means that it is pleasant to human beings as well as to plants, feeling congenial to the skin, neither searingly hot nor depressingly damp. But there are many tropical plants that like to grow in air saturated with moisture and which soon become yellow leaved and unhappy if the air gets too dry. At the other extreme are fleshy leaved and woolly leaved plants that are accustomed to dry air and soon fall victims to decay if grown for long in humid conditions. There is really no way of making plants with extreme requirements of this character happy within the same small greenhouse. Wisdom lies in choosing as companions plants that enjoy similar conditions or, if a wide selection is desired, either providing separate greenhouses for different groups or dividing one greenhouse into several compartments with well constructed partitions to prevent circulation of air from one to the other.

Abutilon, Hibiscus

Abutilons are very attractive plants for the cool greenhouse. Their peak flowering period is during the summer but this can be extended by careful management

These are both fine, shrubby plants, the greenhouse abutilons with hanging, bell-shaped flowers which may be white, yellow, orange or crimson according to variety and *Hibiscus rosa-sinensis* with large, trumpet-shaped flowers which may be scarlet, rose, pink, yellow or buff and are double in some varieties.

Both are best purchased as rooted plants, but they can be easily increased by cuttings of firm young shoots in a propagator in summer or early autumn.

Grow in J.I.P. No. 2 or peat potting compost in the smallest pots that will accommodate the roots comfortably. Quite good specimens can be grown in 6- or 7-in. pots, but old plants may require small tubs. Alternatively, they can be planted permanently in soil beds in the greenhouse.

Water freely in spring and summer, rather sparingly in winter. Shorten overlong or straggly stems each spring. Maintain a minimum winter temperature of 7° C. (45° F.) for abutilons and of 10° C. (50° F.) for hibiscus but no artificial heat should be required from May to September inclusive. Do not shade in summer.

Both abutilons and *Hibiscus rosa-sinensis* have a very long flowering season which is at its peak in the summer but some flowers can be had almost throughout the year if adequate temperatures can be maintained.

Acacia, Mimosa

Flowering in the spring, *Acacia armata* makes an excellent pot plant. It requires little more than frost protection and in summer can stand outdoors in a sunny place

The shrubby plants with fluffy yellow flowers which are commonly called mimosa are in fact acacias and the only true mimosa commonly grown in greenhouses in this country is *Mimosa pudica,* known as the Sensitive Plant because of its strange habit of folding its leaves when touched.

Some acacias make excellent pot plants and will flower in spring in greenhouses with little more than frost protection. The common 'mimosa' of the florists, *Acacia dealbata,* is beautiful but rather too large. *A. armata* and *A. drummondii* make better pot plants.

Grow them in J.I.P. No. 2 or peat-based potting compost and re-pot when necessary, using the smallest pots that will contain the roots comfortably. Good specimens can be grown in 6- to 8-in. pots. Shorten straggly shoots immediately after flowering and water fairly freely in spring and summer, rather sparingly in autumn and winter when a minimum temperature of 7° C. (45° F.) should be maintained. In summer plants may be stood out of doors in a sunny place.

Grow *Mimosa pudica* from seed sown in spring in a temperature of 15° C. (60° F.). Pot seedlings in J.I.P. No. 1 or peat potting compost in 3-in. pots and move on to 5-in. pots when necessary. Water fairly freely and keep in a light, airy greenhouse. It is best to renew from seed annually.

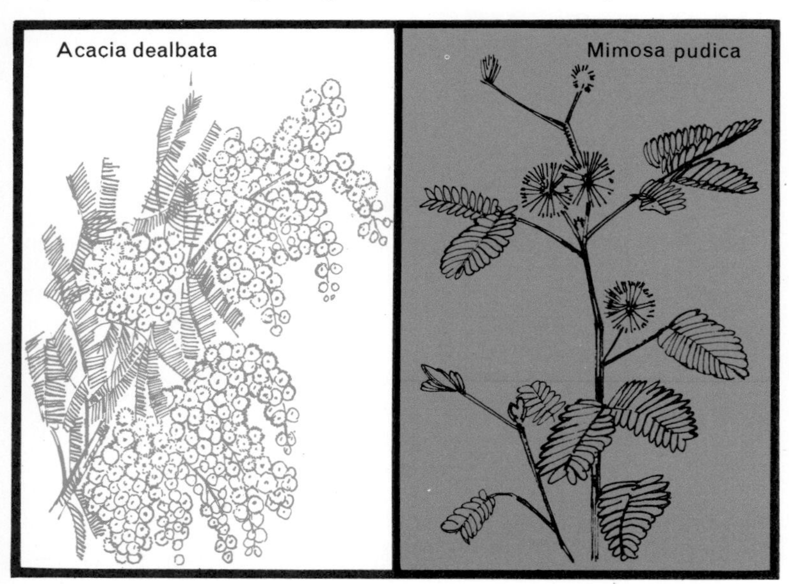

Acalypha and Others

Spectacular variegated foliage is a feature of *Breynia nivosa roseo-picta.* This shrub reaches a height of 3 to 4 ft. and is a marvellous addition to the range of greenhouse foliage plants

The two species of acalypha usually grown require similar treatment but are very different in appearance. *Acalypha hispida* has small magenta flowers produced in summer in long slender tassels, for which reason it is known as Red-hot Cat-tail or Chenille Plant. *A. wilkesiana* has inconspicuous flowers but handsome coppery-green leaves which may be edged with pink, red or yellow in some forms. It is known as Copper Leaf. Both are shrubby plants to be grown in a warm greenhouse in J.I.P. No. 2 or peat-based compost. They need normal watering but a really moist atmosphere in summer with shade. Increase is by summer cuttings.

Asclepias currassavica, known as Blood Flower, is a handsome flowering plant with strong, erect stems terminated by branched clusters of small flowers, purplish-red in bud, red, orange and yellow when open. It flowers for much of the summer and has an all-yellow variety with an even longer flowering season. Asclepias can be grown in J.I.P. No. 2 or a soil border in a cool or intermediate house but should be watered sparingly from November to March. Increase is by seed, division in spring or summer cuttings.

Breynia nivosa, also known as *Phyllanthus nivosus,* is a small shrub. Its variety *roseo-picta* has leaves splashed with pink, red and white. It needs the same treatment as *Asclepias currassavica* and is increased by summer cuttings.

Achimenes and Others

Achimenes are charming plants for hanging baskets. They also make splendid pot plants if the tips of the shoots are pinched out occasionally and some support is provided

Achimenes, columneas and aeschynanthus are related plants with showy flowers all summer and are excellent for hanging baskets. Achimenes are weak stemmed but can be made into bushy, erect plants if the tips of shoots are pinched out occasionally to encourage branching and a few twiggy branchlets are provided for support. Colours are pink, red, purple-blue and white. Columneas and aeschynanthus are trailing and have orange-red flowers.

Achimenes die down in autumn and their little cylindrical roots can be stored dry in a frost-proof place and then restarted in spring in a temperature of 15° C. (60° F.). Columneas and aeschynanthus grow all year and need a minimum winter temperature of 13° C. (55° F.).

Grow all these plants in J.I.P. No. 1 or peat-based compost. Plant five or six tubers of achimenes in each 5-in. pot or place them 2 or 3 in. apart in a hanging basket and cover with 1 in. of compost. One good columnea or aeschynanthus plant is sufficient for a 6-in. pot or hanging basket, but young plants should be started in 3-in. pots and moved on.

Water freely in spring and summer. Allow achimenes to dry off completely in autumn but water the others moderately in autumn and sparingly in winter. Increase achimenes by seed or tubers; columneas and aeschynanthus by cuttings in spring in a well-warmed propagator.

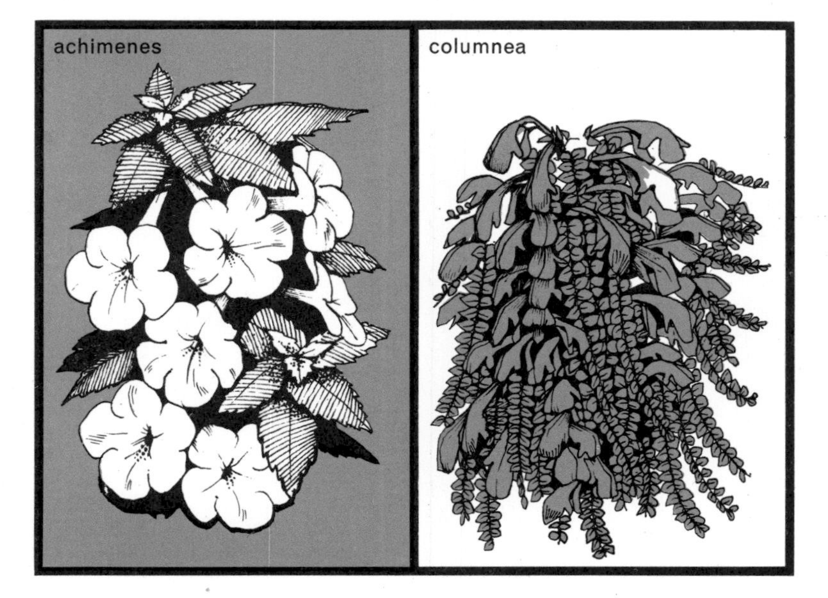

Anthurium, Arum Lily

The highly coloured spathes of *Anthurium andreanum* will look beautiful over a long period. These striking plants demand a humid atmosphere from February through to October

Both these plants carry their true flowers in a spike-like yellow spadix round or behind which is a large spathe, shaped like a shield, pink, scarlet, crimson or white in anthurium, rolled into a funnel and white or yellow in Arum Lily (richardia or zantedeschia). Both flower in spring or early summer.

Grow in J.I.P. No. 1 or peat-based potting compost in the smallest pots that will contain the roots comfortably and pot on as necessary. Quite good plants can be produced in 6-in. pots, but old Arum Lilies may require 7- or 8-in. pots.

Water anthuriums freely in spring and summer and damp down frequently to maintain a moist atmosphere. Shade from direct sunshine. Water and damp down less in autumn and winter and maintain a minimum temperature of 13° C. (55° F.). Re-pot in March and keep the crown of the plant well up above the compost.

Pot Arum Lilies in July or August, water sparingly at first but more freely as growth develops and really freely in spring. Reduce the water supply after flowering and keep the plants nearly dry for a few weeks before re-potting. Maintain a minimum temperature of 7° C. (45° F.). in winter, or 3 to 6° C. (5 to 10° F.) higher if early flowers are required.

Asparagus and Others

An occasional pot plant of *Grevillea robusta* makes a good foil for flowering plants. Individual specimens will soon grow too big but replacements can easily be raised from seed

The kinds of asparagus grown in greenhouses are foliage plants and are often referred to as asparagus ferns, though they are in no way true ferns. The two commonly cultivated are *Asparagus plumosus*, with feathery foliage, and *A. sprengeri*, with narrow leaves and long trailing stems. This is a favourite plant for hanging baskets and *A. plumosus* for foliage to be used with cut flowers.

The grevillea grown in greenhouses, *G. robusta*, is also cultivated for its rather fern-like leaves, but unlike the asparagus, it is a shrubby plant which in time becomes too large for convenience and so is usually renewed every few years from seed. Asparagus can be increased from seed or by dividing the old plants in spring.

Sow seed of both asparagus and grevillea in spring in J.I.S. or peat seed compost in a temperature of 15° C. (60° F.). When the seedlings are about 2 in. high pot them directly into 3-in. pots in J.I.P. No. 1. or peat-based potting compost and move on to larger pots and J.I.P. No. 2 or peat potting compost as necessary. Good plants can be grown in 5-in. pots. Shade from direct sunshine from May to October. Water freely in spring and summer, moderately in autumn and winter, and use artificial heat only to maintain a minimum temperature of 7° C. (45° F.). Train the stems of *Asparagus plumosus* up wires.

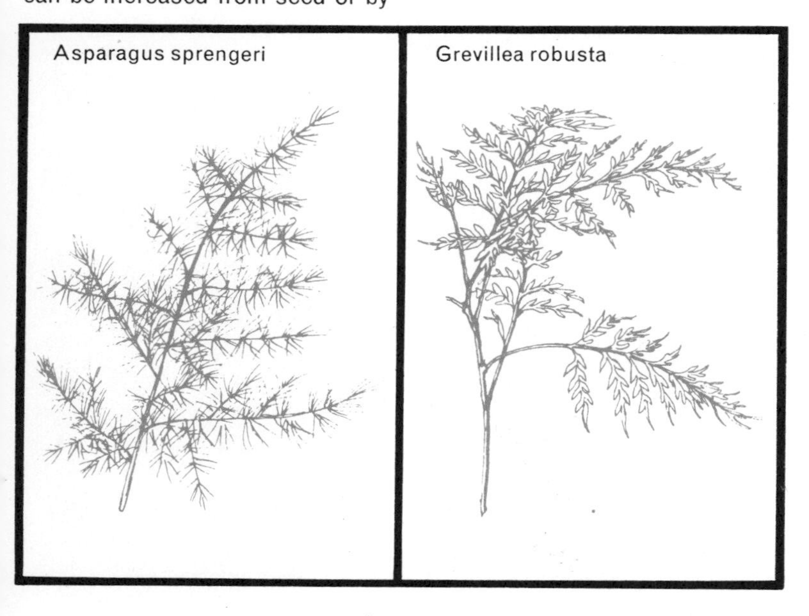

Azaleas

The lovely varieties of the evergreen Indian azalea can be raised in a greenhouse and then removed to the dwelling house when they start to come into flower

The azaleas grown in greenhouses, and very popular for indoor decoration when in flower, are all bushy evergreen shrubs. Their colour range is from white and palest pink to crimson with some salmon and orange-red varieties. The varieties fall into three groups: very early azaleas which can be forced from October to flower at Christmas or early in January; mid-season varieties which can be forced from December to flower from late January to early March, and late varieties which can be forced from January to flower in March or April.

The best way to start is to buy flowering plants and grow them on in a cool greenhouse. Pick off the faded flowers and stand the plants out of doors in a sheltered, partially shaded place. Keep them well watered throughout and from May to September syringe daily with water. Feed every 10 to 14 days from May to August with weak liquid fertiliser. Return the plants to the greenhouse in October before there is a serious frost. For early flowers use early varieties and maintain a minimum temperature of 10 to 13° C. (50 to 55° F.). For later flowers use mid-season varieties and keep in a cool house until December when the temperatures may be raised as above, or use late varieties and grow in cool house conditions throughout. Repot after flowering, using 3 parts peat, 1 part lime-free loam and 1 part sand.

Tuberous-rooted Begonias

The begonia family is a very important one in the greenhouse. Tuberous-rooted begonias make up one group and this variety, Sam Phillips, is an excellent example

The double-flowered varieties of tuberous-rooted begonia make excellent pot plants and the small flowered 'pendula' varieties are equally attractive in baskets. Both can be grown from seed or tubers. The colour range is white, yellow to orange, and pink to crimson.

Sow seeds from January to March in J.I.S. or peat seed compost in a temperature of 15 to 18° C. (60 to 65° F.). Do not cover with soil, only with glass and paper. Prick out as soon as possible into similar compost and later pot singly in 3-in. pots in J.I.P. No. 1 or peat-based compost, moving on as necessary to 5- or 6-in. pots (or baskets) and J.I.P. No. 2.

Alternatively, start tubers in February to March by half burying them in moist peat in a temperature of 16° C. (61° F.). When they have two or three leaves pot as for seedlings.

Grow throughout in a temperature of 15 to 20° C. (60 to 68° F.), shading from sunshine from May to October. Water freely while in growth and feed every 7 to 10 days with weak liquid fertiliser from the time the first flower buds appear. Maintain a cool moist atmosphere throughout and ventilate freely in hot weather. Tie the flower stems individually to short canes. Remove the small female flowers beside the larger male ones. In October gradually reduce the water supply, let growth die down and then cut it off and store tubers dry at not less than 4° C. (40° F.).

planting tubers

Winter Begonias

Winter begonias need a minimum winter temperature of 13°C. (55°F.) but they respond with a wonderful display of flowers during that season. This is the variety Exquisite

The winter-flowering begonias are very beautiful plants but not as easy to grow as the summer-flowering kinds. All can be grown from cuttings of young shoots in spring, rooted in a propagator with a temperature of 18° C. (65° F.).

Pot rooted cuttings in 3-in. pots in J.I.P. No. 1 or peat-based potting compost, re-pot in 4- or 5-in. pots when necessary, and move the largest plants into 6-in. pots for flowering. Water freely and grow on in a minimum temperature of 15° C. (60° F.). Maintain a fairly moist atmosphere and shade from direct sunshine from May to September. In autumn and winter maintain a minimum temperature of 13° C. (55° F.) rising to 18° C. (65° F.) with sun heat. Place several split canes round each plant with encircling ties as support.

After flowering, water sparingly for a few weeks to rest the plants, then cut back the stems, and give more water to restart growth. Take cuttings when the young shoots are sufficiently long.

One of the most popular, Gloire de Lorraine, has fibrous roots, but some other kinds with larger flowers, such as Optima, have semi-tuberous roots. These need specially careful watering in winter since the tubers may decay in too wet a soil. They should not be attempted unless steady winter temperatures of around 15° C. (60° F.) can be maintained.

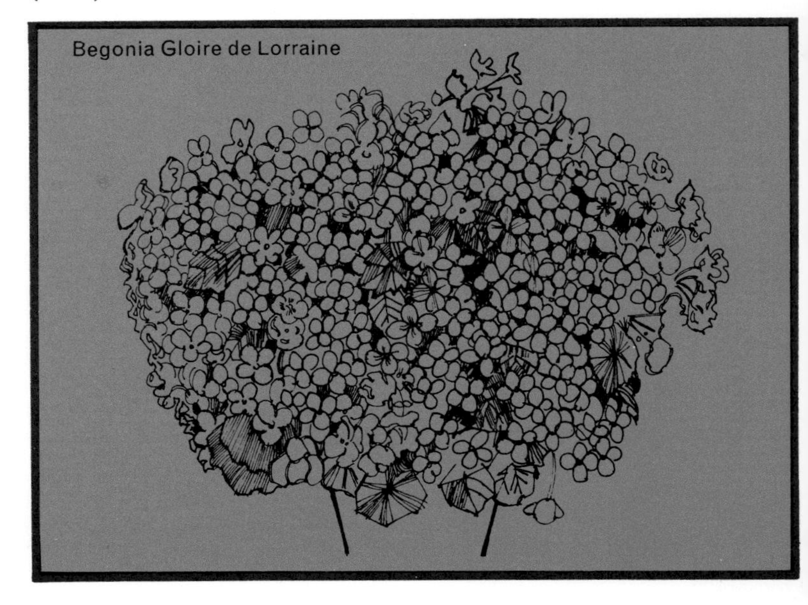
Begonia Gloire de Lorraine

Begonia rex

The large heart-shaped attractively marked leaves of *Begonia rex* make it a handsome foliage plant. Easy to grow, it well deserves a place in any greenhouse

The many varieties of *Begonia rex* are grown for their large heart-shaped leaves which may be green or purple variegated with silver. They are herbaceous plants with thick creeping rhizomes and can be increased by division in spring which is also the best time for planting or re-potting. An alternative method of propagation is to cut off mature leaves in July, make incisions through the leaf veins and then lay them on damp soil or peat and sand in a close moist atmosphere. Plantlets will form from the cut veins and when large enough can be detached and potted. *Begonia rex* can also be grown from seed treated as for tuberous-rooted begonias.

Begonia rex can be grown in a cool greenhouse but is happier in an intermediate house. It should be potted in J.I.P. No. 2 or a peat-based potting compost, or can be planted directly in a bed of fairly rich soil containing some peat. It enjoys shade and can be grown underneath the greenhouse staging. Plants should be watered freely from May to September, fed fortnightly with weak liquid fertiliser, and the atmosphere should be moist. In spring and autumn they should be watered moderately, and rather sparingly from December to February.

Begonia masoniana, sometimes known as Iron Cross because of the shape of the black markings on each green leaf, is grown in the same way.

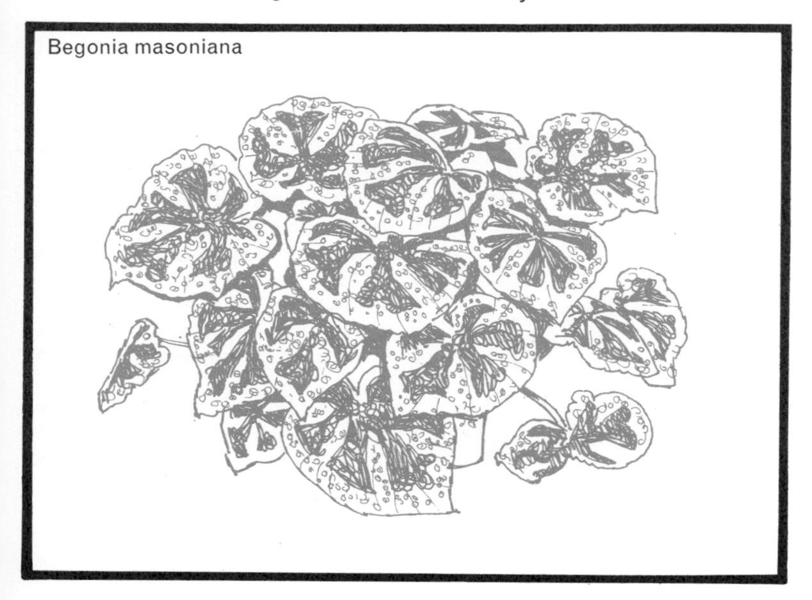

Begonia masoniana

Other Begonias

Begonia Lucerna is a vigorous hybrid capable of growing into a massive ornamental plant. All these begonia species and their hybrids are easy to propagate, particularly by means of cuttings

Begonia haageana

The begonia family is a large one full of good material for the greenhouse. In addition to the types separately described there are numerous species that are easily grown and very decorative. Most also have a long flowering season. *Begonia coccinea* is 5 to 6 ft. high with pendant sprays of scarlet flowers from May to October. *B. evansiana* is shorter and has small pink flowers all summer. *B. fuchsioides* makes long slender stems which can be trained against a wall. Its sprays of small pink flowers are produced from autumn until late spring. *B. haageana* has large leaves green above and purple beneath and pink flowers in summer. Lucerna and President Carnot are good hybrids in a similar style. *B. sutherlandii* is a foot high and has orange flowers all summer. There are many more.

All these can be grown in pots in J.I.P. No. 2 or peat potting compost in an intermediate greenhouse, or the summer-flowering kinds (particularly *B. evansiana* and *B. sutherlandii* which are nearly hardy) in a cool house. The summer-flowering kinds should have normal watering; the winter-flowering kinds should be rested and kept a little dry for a few weeks after flowering. All should be shaded and given a moist atmosphere in summer.

All species can be raised from seed as described on page 62. Some can be increased by division in spring and most by cuttings of non-flowering shoots in summer.

Beloperone, Jacobinia

Beloperone guttata is called the Shrimp Plant because of its sheath-like pink bracts. It will flower all year if a minimum temperature of 10°C. (50°F.) can be maintained

Beloperone guttata is known as the Shrimp Plant because of the sheath-like pink bracts which cover the nodding flower spikes. There is a greenish-yellow variety and also a dwarf variety only about 12 in. high against a normal 18 to 24 in. All flower almost throughout the year if a minimum temperature of 10° C. (50° F.) can be maintained. They will grow well in J.I.P. No. 1 or equivalent compost, and will survive in a cool greenhouse (though may not then flower all winter for which an intermediate house is necessary). Water normally. Increase by cuttings.

Jacobinias are of two distinct types, one with close heads of flowers at the tops of the stems, the other with tubular flowers scattered singly all over the plant. *Jacobinia carnea* (or *Justicia carnea*) with pink flowers in August and September is the most popular of the first type. It will grow to 6 ft. but can be kept at 3 ft. if it is fairly hard pruned after flowering. *J. pauciflora* (or *Libonia floribunda*), with scarlet and orange flowers from autumn until summer, is the most popular of the second type.

Both plants can be grown in J.I.P. No. 2 in an intermediate house. *J. pauciflora* can be put outdoors or in a frame from June to September. Water both normally except that *J. carnea* can be kept a little dry for a few weeks after pruning. In the greenhouse in summer they need light shading. Increase by cuttings of firm young growth.

Billbergia and Relatives

The bromeliads are a family of curious rosette plants. Many are remarkable for the colour and variegations of their leaves as well as for their flowers, as shown above in *Billbergia nutans*

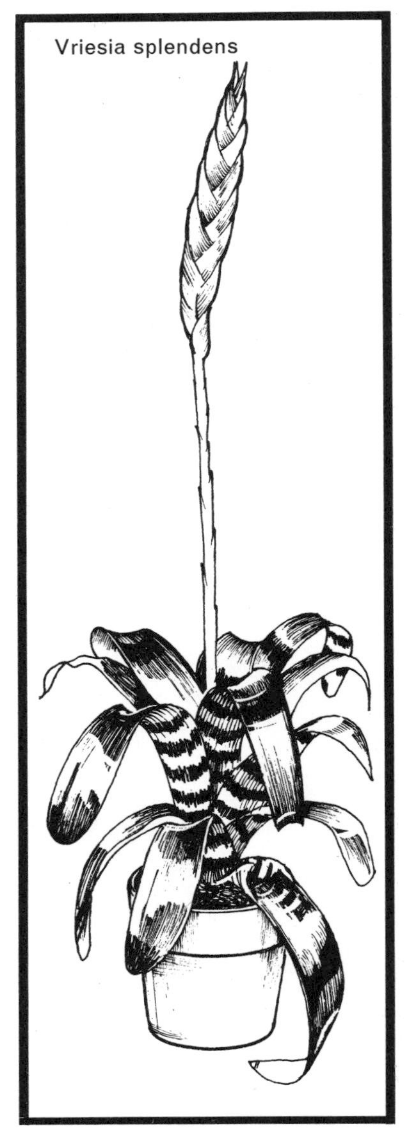

Billbergia nutans belongs to the bromeliads. It makes clumps of rather stiff narrow saw-edged leaves and from June to August has slender 12- to 18-in. arching flower stems with drooping tubular green and blue flowers surrounded by pink bracts. It is quite easy to grow in a cool or intermediate greenhouse in J.I.P. No. 1, requires light shading from June to October and very little water in winter. Water normally during the rest of the year and feed occasionally in summer.

Many other popular bromeliads, such as aechmea, nidularium and vriesia, have strap-shaped leaves forming stiff rosettes with a hollow, often called a 'vase', in the middle. These leaves may be mottled or grey-green or be handsomely variegated and in some, e.g. *Neoregelia carolinae*, are scarlet around the vase. Others are handsome in flower – *Aechmea fasciata* with heads of blue flowers and pink bracts, and *Vriesia splendens*, with spikes of scarlet bracts. Cryptanthus carries its curious bronzy leaves in rosettes without a vase.

All these like an intermediate or warm house, well shaded and humid in summer. Most should be watered normally except that the central vase, if present, should always be full of water. Keep cryptanthus rather dry in winter. All can be grown in equal parts of sand, peat and osmunda fibre in the smallest pots that will contain them. Increase is by offsets removed in spring.

Boronia and Others

Bouvardia is a beautiful evergreen shrub with lovely waxy flowers. These are produced during the autumn and winter and are available in white, pink and red

Boronias are small evergreen shrubs, the most popular of which, *Boronia megastigma*, has very narrow leaves like those of a heath and maroon and yellow flowers from February to May. Others grown are *B. elatior*, carmine and *B. heterophylla*, purple. They can be grown in a cool greenhouse without shading and can be stood out of doors in a sunny place from June to September. They should be grown in J.I.P. No. 1 with some extra peat and must be watered carefully as they resent very wet soil, particularly in winter. Increase is by cuttings in summer in a propagator or under mist.

Bouvardias are also evergreen shrubs. The brightly coloured tubular flowers are freely produced in terminal clusters on 2- or 3-ft. stems from October to February and may be single or double, pink, scarlet or white. Plants should be grown in J.I.P. No. 2 and kept in an intermediate house, lightly shaded in summer or stood outdoors in a sheltered place from June to September. They are watered normally but after flowering they should be cut back and watered sparingly for three or four weeks. *Bouvardia longiflora*, which has white scented flowers, can be grown fairly successfully in a cool house.

Pentas lanceolata resembles bouvardia in appearance and requires similar treatment. It has rosy purple, pink or white flowers in autumn and winter.

All can be increased by spring cuttings in a propagator.

Browallia and Others

Celosias are half-hardy annuals which can be grown as pot plants or raised in the greenhouse for bedding out in summer. There are two types, one of which, *Celosia plumosa*, is illustrated

Browallia speciosa is an attractive annual that will flower in a cool or intermediate greenhouse from July to December if two sowings are made. The first should be in February or March, the second in May or June. Raise the seedlings in J.I.S. compost and transfer them singly to 4-in. pots in J.I.P. No. 1. Water fairly freely throughout and do not shade. Good varieties are Major, blue, Silver Stars, white and Sapphire, blue. All are about 1 ft.

Celosias are also annuals. Two quite distinct types are grown, *C. plumosa*, the Prince of Wales' Feather, with plumy sprays of yellow, orange, red or cerise flowers, and *C. cristata*, the Cockscomb, with flowers in similar colours crowded into flat, twisted heads like a cock's comb. Both are summer flowering and make excellent pot plants for a cool greenhouse. Seed should be sown from February to April and seedlings treated as for browallia. Plants can be stood outdoors in a sunny sheltered place in summer.

Celsia arcturus looks like a small verbascum or mullein, to which it is related. It is a half-hardy herbaceous plant readily raised from seed which can be sown in March to give plants to flower in September–October or in June–July to flower from April to July or later. Seedlings should be raised in the same way as browallia and if to be over-wintered should be kept in a cool greenhouse. It is often discarded after flowering.

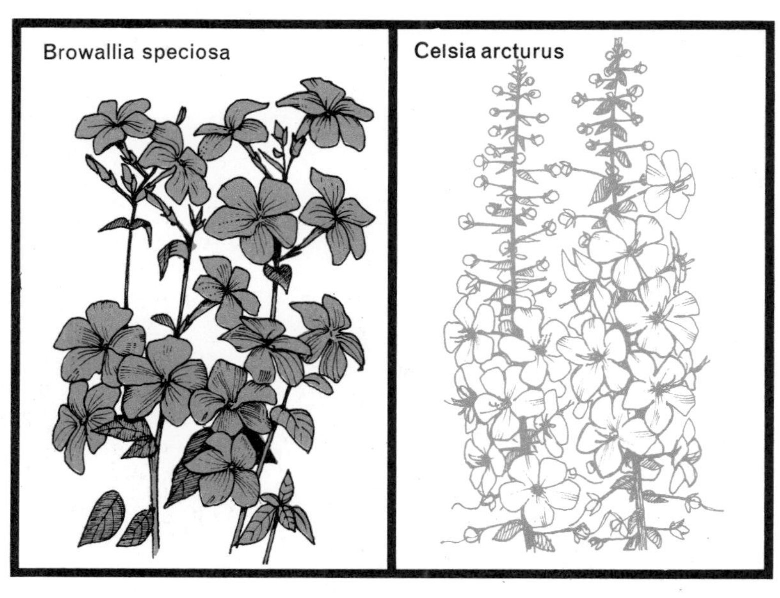

Brunfelsia, Gardenia

Brunfelsia calycina, a native of Peru, is a rewarding plant to grow provided a minimum winter temperature of 10°C. (50°F.) can be maintained. The fragrant flowers are produced for most of the year

Brunfelsias are low-growing evergreen shrubs with showy flowers produced most of the year. They can be grown in a cool house, but are happier and flower more continuously in an intermediate house. They can be grown in large pots in J.I.P. No. 2 or peat-based compost or be planted directly in a bed of good loamy soil. They should be shaded in summer and kept in a moist atmosphere. Water freely from April to October and feed occasionally from May to July with weak liquid fertiliser. Water rather sparingly from November to March but do not allow plants to become really dry. Prune lightly each spring. Increase by cuttings of firm young shoots in June and July in a propagator. *Brunfelsia calycina*, with violet-blue flowers, is the most popular species.

Gardenia jasminoides, the Cape Jasmine, is an evergreen shrub requiring similar treatment and also capable of flowering on and off throughout the year, especially in intermediate or warm greenhouses. It should be grown in J.I.P. No. 2 with normal watering. Shade lightly in summer if under glass, but plants can be stood out of doors from June to September in a sunny sheltered place. Increase by cuttings of young shoots in March and April in a propagator. Most of the varieties cultivated have double flowers. All are very sweetly scented. *G. j. veitchiana* is specially recommended for winter flowering.

Caladium, Codiaeum

Caladiums are among the most beautiful of greenhouse foliage plants. They have very attractive large shield-shaped leaves marked with green, white, pink and rose

Both these plants are grown solely for their highly decorative foliage. The leaves of caladiums are large and shield-shaped, green, white, pink or rose, variously veined, mottled and edged with one colour on another. Codiaeums, more familiar to many gardeners by their former name, croton, are bushier in habit and have evergreen, lance-shaped or oval leaves beautifully mottled in green, yellow, orange, crimson and bronze.

Caladiums have tuberous roots which should be potted in March in either peat-based potting compost or J.I.P. No. 1 with an additional 25 per cent. of peat. Water sparingly at first, freely as growth starts, and grow in a temperature of 18 to 24° C. (65 to 75° F.) throughout, maintaining a moist atmosphere. Shade from strong sunshine but grow in good light as this helps to develop the full colour of the leaves. In autumn the water supply should be greatly reduced and in winter the plants should be kept in a minimum 13° C. (55° F.).

Pot codiaeums in spring in J.I.P. No. 2 or peat-based potting compost and grow on under the same conditions as caladiums. Do not dry off in autumn but water moderately from then until spring and maintain a temperature of about 15° C. (60° F.) or a minimum of 13° C (55° F.).

Increase caladiums by division at potting time; codiaeums by cuttings of firm young shoots in a warm propagator in summer.

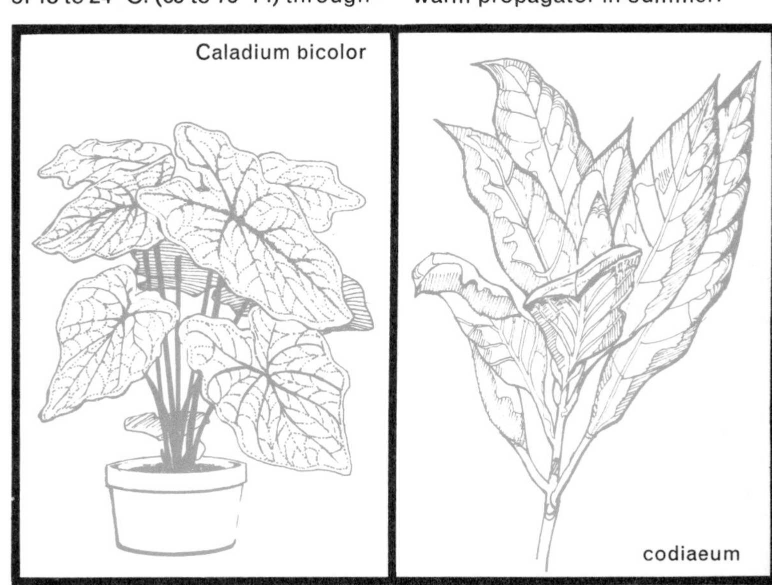

Calceolarias

The gaily coloured, pouched flowers of calceolarias are often spotted with contrasting hues. The varieties differ in height and all make excellent pot plants

The greenhouse calceolarias all have large pouched flowers in a variety of colours including yellow, orange, red and crimson, often with one colour brilliantly spotted or splashed on another. There are both tall and dwarf varieties, the former 15 to 18 in. high, the latter 9 to 12 in. and all make first-class pot plants to flower in a cool greenhouse in May and June. They are grown as biennials to be renewed from seed annually.

Sow seed in J.I.S. or peat-based seed compost in an unheated greenhouse or frame in May or June. Prick out the seedlings into boxes of similar compost and when they have formed a few leaves each pot them singly in 3-in. pots in J.I.P. No. 1 or peat-based potting compost. They should be shaded from direct sunshine, watered fairly freely and given ample ventilation. Indeed, since they are nearly hardy plants, at this stage and until late September they are really better in a frame than in a greenhouse.

They should be potted on into 4-in. pots and, if necessary, once again into 5-in. pots, in J.I.P. No. 2 or peat potting compost, and after September should be kept in a light greenhouse with a minimum temperature of 7° C. (45° F.). Water rather sparingly in winter but give more water in spring. The taller plants will need careful staking.

Callistemon, Clerodendrum

Clerodendrum thomsoniae is a striking evergreen climber with clusters of flowers during the summer and autumn. It looks attractive when trained over a crinoline frame

Callistemons are evergreen Australian shrubs with dense spikes of pink, red or yellow flowers. They make excellent pot plants grown in J.I.P. No. 2 compost in a sunny greenhouse with no more than frost protection in winter. From May to October the plants can stand out of doors; in mild places they can be planted permanently out of doors. Good kinds are: *Callistemon coccineus*, red and yellow, *C. lanceolatus* and *C. speciosus*, scarlet or crimson and *C. salignus*, pale yellow.

Two very different kinds of clerodendrum are grown in greenhouses, *Clerodendrum fallax*, a shrubby plant with handsome heads of scarlet flowers in summer, and *C. thomsoniae*, an evergreen climber with clusters of red and white flowers in summer and autumn. Both can be grown in pots, or *C. thomsoniae* can be planted in a bed and trained up a greenhouse wall or over a crinoline frame in a minimum 10° C. (50° F.).

Grow in John Innes No. 1 or peat potting compost. Pot in spring in the smallest pots that will contain the roots and re-pot as necessary. Water freely in spring and summer, sparingly in autumn and winter and prune after flowering. Cuttings may be prepared then and rooted in a propagator at 18 to 21° C. (65 to 70° F.). *Clerodendrum fallax* can also be raised from seed sown in spring in 18 to 21° C. (65 to 70° F.).

Campanula, Exacum

Exacum affine, a delightful greenhouse annual, can be seen in flower throughout the summer, autumn and winter depending on the time of sowing. The flowers are very fragrant

Two very different campanulas are commonly grown in greenhouses. One is *Campanula isophylla*, a trailing perennial with star-shaped lavender or white flowers in August and September. It is an excellent plant for hanging baskets and is easily grown in J.I.P. No. 1 or peat-based potting compost, in sun or shade. It should be watered normally and can be increased by division in spring.

Campanula pyramidalis is known as the Chimney Bellflower because of its narrow 4- to 5-ft. spikes of blue or white flowers in June and July. It is grown as a biennial from seed sown in March or April in a cool or unheated greenhouse. Seedlings are raised in J.I.S. compost, potted singly into J.I.P. No. 1 and later moved on to larger pots and J.I.P. No. 2. Pots up to a 10 in. diameter may be required eventually. Plants can stand outdoors from June to September. Watering is normal throughout.

Exacum affine is a bushy annual 9 to 18 in. high with lilac-blue and yellow, scented flowers freely produced from August to December according to date of sowing. Two sowings are usually made, one in March, the second in June, in J.I.S. compost and a temperature of 15° C. (60° F.). Seedlings are potted singly in J.I.P. No. 1 and may later be transferred to 4-in. pots and J.I.P. No. 2. They should be watered fairly freely, should be grown in a cold or cool house with good light and a fairly damp atmosphere.

Canna, Hedychium

Hedychium spicatum acuminatum is a member of the Ginger family. It reaches a height of 3 ft. and is grown in the same way as the other species described below

The canna or Indian Shot is a tropical plant with broadly lance-shaped, green or purple leaves and showy spikes of summer flowers, red, yellow, orange or pink, often blotched with one colour on another. It has fleshy roots which can be stored dry in winter and can be grown on with very little artificial heat. Roots should be potted in February, March or April in J.I.P. No. 2 or peat-based potting compost and started into growth at 15 to 18° C. (60 to 65° F.). Thereafter they should be watered freely, fed every 10 to 14 days from June onwards with weak liquid fertiliser and given plenty of light and sunshine. In October water can be gradually withheld and when growth has died down the roots, in their soil, can be stored dry in any frost-proof place. Plants can be grown from seed but this should be soaked for 24 hours then nicked and germinated at 20° C. (68° F.).

Hedychium gardnerianum has similar, but always green, foliage and short broad spikes of pale yellow, scented flowers in summer. It, too, looks tropical but can be grown in a cool greenhouse if it is kept nearly dry all winter. In an intermediate house it can be given more water and will keep on growing, and this treatment also suits orange-red *H. coccineum* and white *H. coronarium*. Otherwise they are grown like cannas, except that they are best planted in a bed of good soil. *H. gardnerianum* is from 4 to 6 ft. the others 3 to 4 ft.

Capsicums

Capsicums are grown both for their decorative value and their culinary uses. They are easily raised from seed and can be grown in a cool greenhouse or a sheltered place outdoors

In addition to the culinary varieties of capsicum grown for their edible fruits there are a number of ornamental varieties differing in the shape and colour of their fruits. These sweet peppers or chillies can make highly decorative pot plants and are also useful for floral arrangements. They make bushy plants usually with the fruit carried above the foliage, and these fruits may be oval, conical or pencil shaped, erect or pendulous, straight, curved or twisted, green, yellow or red.

All are grown as annuals from seed sown from January to March in a temperature of 18° C. (65° F.). Raise the seedlings in J.I.S. compost, prick them out in J.I.P. No. 1 and pot singly in J.I.P. No. 2. Grow on in a sunny greenhouse, temperature 13 to 18° C. (55 to 65° F.), watering freely. Plants can be hardened off in a frame in May for planting out of doors in early June in good soil and a warm sunny place. Or they may be grown on in pots, 4 to 5 in. in diameter for the popular dwarf 8- to 12-in. varieties, or up to 7 in. diameter for the tall 2- to 3-ft. varieties. They can either be kept in a sunny, well ventilated greenhouse or in a frame and must be well watered and fed every 10 to 14 days with weak liquid fertiliser from the time the first fruits are set. Plants in flower should be syringed daily with water to assist setting of the fruits. Tall varieties must be staked and tied. The ripe fruits retain their colour for a considerable time.

Cinerarias

Cinerarias are easily grown from seed and they are excellent subjects for the cool greenhouse. Flowering during the winter and spring they bring welcome colour at this time

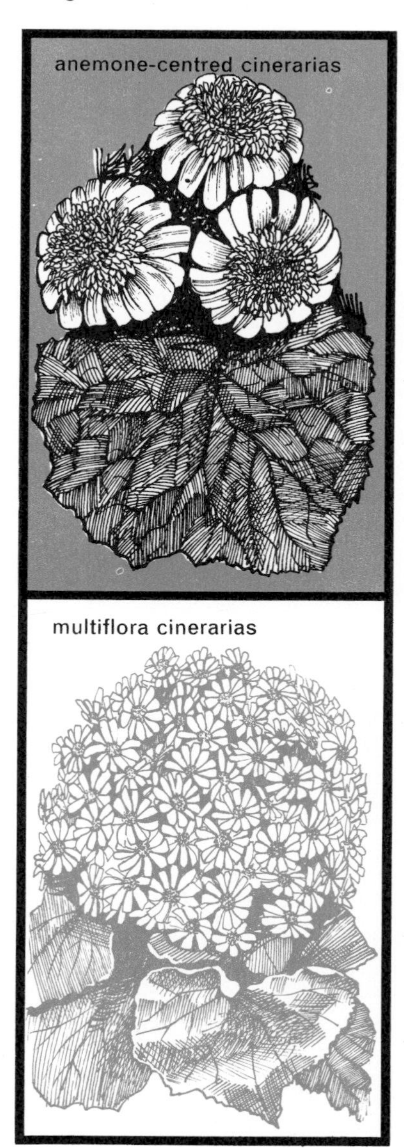

Cinerarias are herbaceous perennials but are almost invariably grown from seed as biennials, being discarded after flowering. Only double-flowered varieties which set no seed are grown from cuttings and nowadays these are rarely seen.

The daisy-like flowers in a range of rich colours including shades of blue, violet, purple, crimson and scarlet, often with a white zone around the central disc, are produced in fine heads from November to May according to the time of sowing and the temperature available. There are Stellata varieties with small flowers, Multiflora varieties with flowers of medium size and large-flowered Grandiflora varieties. There are also dwarf varieties 12 to 15 in. high compared with the normal 18 to 24 in.

Seed should be sown in April, May and June for the longest flowering season. Sow in J.I.S. or peat-based compost in a temperature of 13 to 15° C. (55 to 60° F.). Prick out seedlings into similar compost and later pot singly in 3-in. pots in J.I.P. No. 1 or peat-based potting compost. Grow on in a light, unheated greenhouse or frame with shade from direct sunshine only. From late September onwards keep plants in a cool unshaded house. Water fairly freely in summer, sparingly in winter and ventilate as much as possible whilst maintaining a minimum temperature of 8° C. (46° F.). In a stuffy, warm, damp atmosphere cinerarias are liable to rot at soil level.

Perpetual Carnations

There are many kinds of perpetual-flowering carnations and new ones are introduced every year. Peppermint Sim, shown above, originated in America

Zuni is another variety of the perpetual-flowering carnation. These plants are grown for their importance as cut flowers rather than for any decorative value in the greenhouse

Provided that a minimum winter temperature of 7° C. (45° F.) can be maintained, perpetual-flowering carnations will bloom all year. They enjoy cool, airy conditions and dislike damp, which encourages stem diseases, and excessive heat, which can produce serious attacks by thrips and red spider.

Perpetual carnations can be grown from seed but the quality and colour of the flowers is very variable and the more usual way is to start with young plants of named varieties and renew these at least every second year by cuttings. These are prepared from November to March from non-flowering side growths that appear midway up the flowering stems. These sideshoots are broken out when about 3 in. long and are rooted in sand in a glass-covered box or frame with just a little bottom heat. When well rooted they are potted singly in 3-in. pots in J.I.P. No. 1 compost and are moved on as necessary, first to 5- or 6-in. pots and later to 7- or 8-in. pots. Young plants are best purchased in spring and potted into J.I.P. No. 2.

Plants should be watered moderately throughout, just sufficient being given to keep the soil moist.

Break out the top two joints of each plant when it has made eight pairs of leaves and repeat the process on the sideshoots that result from this first stopping when they have produced six pairs of leaves. Do not stop after the first week of June if winter flowers are required.

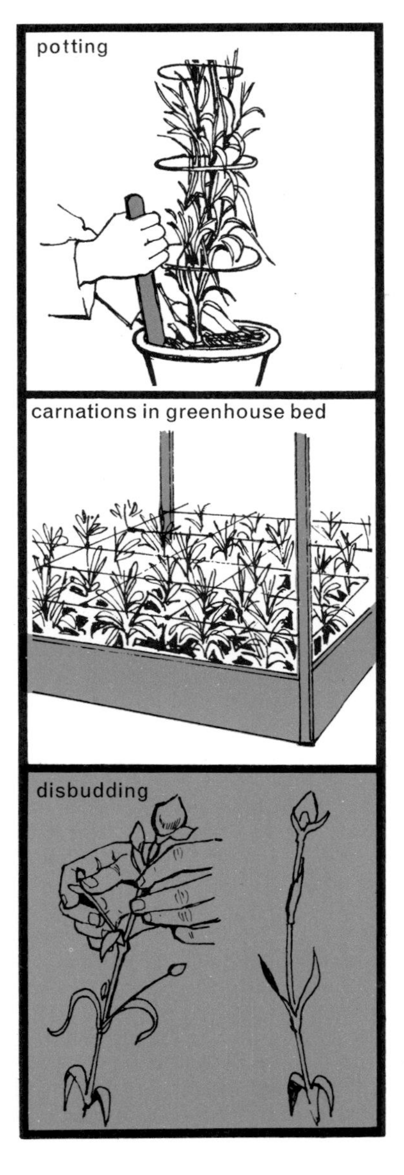

Each plant must be staked, preferably with special wire carnation supports which are readily adjustable as the plants grow. Two-year-old plants may be 4 or 5 ft. high. Keep the plants in a light, well ventilated house without any artificial heat from May to October. Alternatively, from June to September the pots may be plunged to their rims in peat or sand in a sunny, sheltered place outdoors. No shading is required at any time. Framed plants should be returned to the greenhouse before there is frost, and during autumn and winter all should be kept at a temperature of 10 to 15° C. (50 to 60° F.) which can fall occasionally to 7° C. (45° F.).

When the plants form flower buds carefully remove the small side buds and retain only the topmost bud on each stem. This disbudding is not necessary with some rather small-flowered kinds known as Spray carnations.

The greenhouse should be fumigated occasionally, especially from May to August, to destroy thrips and red spider mites. It is unwise to attempt to keep plants after their second year as they usually become weak.

There are a great many varieties in a wide colour range, but perpetual carnations lack fragrance. This is found in the Malmaison carnations which are grown in a similar manner but only flower in spring and summer. As a rule they are increased by layering in July and are only stopped once.

Citrus, Avocado

The Calamondin Orange, *Citrus mitis*, will add interest to any cool greenhouse. To encourage the formation of flowers and fruits it is important that plants receive maximum sunshine in summer

Citrus is the name for all the oranges, lemons and grapefruits, any of which can be grown as greenhouse plants. But the best, as it is the neatest in habit and usually the most prolific, is the Calamondin Orange, *Citrus mitis*. Like other kinds this has shiny evergreen leaves and white fragrant flowers in spring and early summer. The fruits are small and resemble those of a tangerine.

All citrus varieties should be grown in a cool house and may be placed outdoors in a sunny sheltered place from June to September. They can be grown in large pots or tubs in J.I.P. No. 2 compost. Water freely in spring and summer, moderately in autumn, sparingly in winter. No shading is required and sunshine is essential to ripen the growth and ensure flowering and fruiting. Overgrown or badly shaped plants can be pruned lightly after flowering when it can be seen where the fruits have set. Increase is by summer cuttings of firm young growth in a propagator, or from pips.

Persea americana is the Avocado Pear, a large evergreen tree with handsome foliage which is only suitable for pot culture as a young plant. It is raised from seeds which can be obtained from purchased avocado pears. These germinate readily in J.I.S. or peat-based seed compost in a temperature of 15 to 18° C. (60 to 65° F.) and can be grown on in J.I.P. No. 2 in a cool or intermediate house or a sunny window until they get too large.

Clianthus, Erythrina

Clianthus looks very attractive when sprawled over a support or another plant. The striking flowers have given rise to the descriptive common names of Lobster Claw or Parrot's Bill

Two very different species of clianthus can be grown in the greenhouse. *Clianthus puniceus*, the Lobster Claw or Parrot's Bill, is a vigorous shrub, usually evergreen under glass, with long slender stems which are best tied out on a trellis or wires. The scarlet or pure white flowers are shaped like a lobster's claw and appear in May and June. The Glory Pea, *C. dampieri*, is a more sprawling plant with greyish leaves and scarlet flowers, each with a prominent black blotch. It is difficult to grow on its own roots and is often grafted on to *Colutea arborescens*.

Both species can be grown in a cool greenhouse in J.I.P. No. 1. *C. puniceus* is really best planted in a bed of good loamy soil, but it can be grown in large pots or tubs. It can be raised from seed sown in spring or from summer cuttings of firm young shoots in a propagator. Both kinds need much sunshine and should be watered freely from April to October, sparingly from November to March.

Erythrina crista-galli, the Coral Tree, requires similar conditions and treatment and its crimson flowers, borne in late summer in long spikes, also resemble lobster claws. It is a shrub with fleshy roots but it can be cut hard back each March and will then make arching growths 4 to 6 ft. in length. Keep almost completely dry in winter and at all times give all the light and sunshine possible. It can be fed fortnightly with weak liquid fertiliser from May to August.

Coleus

Coleus, with their handsomely coloured leaves, are splendid plants for the amateur's greenhouse but grow best in a minimum winter temperature of about 13°C. (55°F.)

One of the most useful foliage plants for the greenhouse is the coleus with its nettle-shaped leaves in various colours, often with one colour splashed or zoned on another. As a rule mixed colours only are offered in seed though it is possible to produce varieties that will come reasonably true to colour. More usually these specially selected colours are reproduced by cuttings.

Seed is sown in March or April in J.I.S. or a peat-based seed compost in a temperature of 18° C. (65° F.). Seedlings are transferred to 3-in. pots in J.I.P. No. 1 or a peat-based compost and are later moved on into 5-in. pots in similar compost. The tip of each plant is pinched out when it has made four pairs of leaves. Alternatively, plants can be restricted to a single stem until they are 12 to 18 in. high and then pinched to form little standards. Either way, sideshoots are again pinched at four leaves.

Cuttings of young sideshoots root readily in late summer in a close frame or propagator.

Grow throughout in a cool or intermediate house, the latter being preferable if plants are to be overwintered as they are readily attacked by grey mould at temperatures below 13° C. (55° F.). Water freely from April to September, sparingly for the rest of the year and shade lightly from June to September. They can be fed with weak liquid fertiliser from June to August.

Correa, Chorizema

The small pea-like flowers of *Chorizema cordatum* show a brilliant combination of colours. This Australian shrub will reach a height of 5 ft. but it resents any root disturbance

The correas are attractive little Australian or Tasmanian evergreen shrubs with pendant tubular flowers freely produced for many months from late winter onwards. In Australia they are known as Native Fuchsia. They are sufficiently hardy to be grown outdoors in the mildest parts of Britain, but are more reliable and also earlier flowering in a cool greenhouse. They can either be grown in fairly large pots in J.I.P. No. 2 compost, with a little extra sand as they appreciate good drainage, or be planted in a border of loam with which plenty of peat and sand has been mixed. Water moderately in spring and summer, less in autumn and winter. Light shading is beneficial. If they get too large they can be lightly pruned in late spring. Increase is by summer cuttings of firm young growth. The kinds most likely to be available are *Correa alba*, creamy white; *C. backhousiana*, cream tinged with green; *C. pulchella*, pink and red, and *C. reflexa* bronze red and yellow.

The chorizemas, or Flame Peas, also from Australia, are evergreen shrubs and require similar conditions to correas with even greater emphasis on sharp drainage and careful winter watering. They have small, brightly coloured pea-like flowers very freely produced in spring and summer. They can be raised from summer cuttings of firm young shoots, which should be potted singly as soon as rooted and then grown on without root breakage which can cause death.

Crossandra, Aphelandra

Silver Queen is a lovely variety of *Aphelandra squarrosa louisae*. Aphelandras grow best in a greenhouse but can be taken into the dwelling house for short periods

Aphelandra squarrosa louisae has already been mentioned as an evergreen house plant. In the diminished lighting of a room, however, it seldom flowers well and is valued chiefly as a foliage plant. In an intermediate or warm greenhouse, with only light shading in summer from direct sunshine and plenty of moisture in the air from May to September, it will produce plenty of its very handsome flower spikes in summer. Each spike is composed of close packed yellow flowers that look almost as if they are carved out of wood. Aphelandra should be grown in J.I.P. No. 2 or peat-based compost. Plants should be watered freely in spring and summer but sparingly in autumn and winter. They can be fed with weak liquid fertiliser every fortnight in summer. After flowering, stems can be shortened to keep plants compact and bushy. Increase is by cuttings of young growth in a propagator.

Crossandras are related to aphelandras and require similar conditions but if anything slightly more warmth, which makes them happier in the warm rather than the intermediate greenhouse. They, too, produce their showy flowers in short, close packed, terminal spikes. *Crossandra infundibuliformis* (*undulifolia*), with orange flowers in spring, is the kind most usually seen. It can be raised from seed sown in a temperature of 18 to 21° C. (65 to 70° F.), or from cuttings as for aphelandra.

Crossandra infundibuliformis

Cuphea, Manettia

Cuphea ignea comes from Mexico and is of interest for the delicate, slightly charred-looking flowers, which appear over a long season and have given rise to the common name of Cigar Flower

Manettia inflata

Cuphea ignea is known as the Cigar Flower because its little scarlet tubular flowers are black and white at the tip as if charred. They are produced continuously from spring to autumn. It is a bushy perennial about 1 ft. high, sufficiently hardy to grow outdoors from June to September. For the rest of the year it requires cool house treatment. It is easily raised from seed sown in a temperature of 15° C. (60° F.) in March or from spring cuttings of young shoots in a propagator. It should be watered normally and can be grown in 4-in. pots in J.I.P. No. 1 or peat-based compost. *C. microphylla*, with yellow and red flowers, is also grown.

Manettias are slender climbers also with small tubular flowers produced for much of the year. *Manettia bicolor* and *M. inflata* are most commonly seen, both with red and yellow flowers, but more yellow in the former and red in the latter. They need more warmth than the cupheas and should be grown throughout in an intermediate or warm house with plenty of humidity and shade from direct sunshine from May to October. They will thrive in J.I.P. No. 1 or peat-based compost, or can be planted in a border of good loam with some peat and sand. Canes should be provided for the slender stems to twine around. Water normally, feed with weak liquid fertiliser every 14 days from May to August and prune as needed in February. Increase is by cuttings in May or June in a propagator.

Cyclamen

One of the outstanding greenhouse plants, cyclamen need careful treatment throughout their culture. Given the correct conditions, however, they will present a beautiful display

The lovely butterfly-like flowers of the cyclamen are produced from autumn to spring, and particularly attractive are the varieties with heart-shaped leaves heavily marbled with silver.

Sow seed in August in J.I.S. or peat-based seed compost in an unheated greenhouse or frame. Germination may be irregular so carefully lift out the seedlings as they reach the two-leaf stage and prick them out in similar compost. Keep them in a greenhouse all winter, minimum temperature 7° C. (45° F.), watering rather sparingly, and in March or April pot singly in 3-in. pots in J.I.P. No. 1 or a peat potting compost. At this stage maintain a minimum temperature of 10° C. (50° F.) rising to 15 to 18° C. (60 to 65° F.) with sun heat, and water fairly freely. In June, pot on into 5-in pots in J.I.P. No. 2 or peat potting compost and grow on, ventilating freely and shading from strong, direct sunlight only. The plants may be kept in a frame from June to September. Return them to the greenhouse in late September and maintain a temperature of 13° C. (55° F.) if early flowers are required.

In spring after flowering gradually reduce water until the leaves die down, then keep almost dry in a frame. In July re-pot, keeping the corms almost on top of the soil. Water more freely and return them to the greenhouse in September. Cyclamen may be kept thriving for many years.

Cyperus, Cycas

The feathery foliage of *Cycas revoluta* makes it a very decorative plant for use in large greenhouses where a suitable range of temperature can be maintained

Cyperus alternifolius is known as the Umbrella Grass because of its bare stems each bearing at the tip a circle of narrow green leaves arranged like the ribs of an umbrella. It is usually about 18 in. high and is a distinctive foliage plant for cool greenhouses. There is a variety with white variegated leaves. Both can be grown in J.I.P. No. 1 compost in pots which should either be well watered in spring and summer or may be stood nearly to their rims in a pool. In summer the Umbrella Grass is quite happy out of doors. *C. papyrus*, the Egyptian Paper Reed or Papyrus is more tender and larger. It can easily reach 8 ft. and the green leaves are more numerous, much narrower and drooping. Treatment is similar to that for *C. alternifolius* except that an intermediate or warm house is required. Increase is by division in spring.

All species of cycas require a lot of room and the temperature of an intermediate or warm house. Apart from that they are not difficult to grow in J.I.P. No. 2 with normal watering and shade, plus a very humid atmosphere from May to September. Plants sometimes form suckers which can be carefully detached with roots in spring and potted. Alternatively, plants can be increased by seeds sown in spring in a temperature of 26 to 30° C. (80 to 85° F.). The kinds most often seen are *Cycas circinalis*, with leaves up to 10 ft. long and *C. revoluta*, the Sago Palm, with leaves 2 to 3 ft. long.

Datura

Datura sanguinea is a large shrub growing from 4 to 8 ft. high and producing its spectacular flowers, each of which may be up to 8 in. long, in summer

Three species of datura, all shrubby with large white trumpet-shaped hanging flowers, are known to gardeners by the same popular name, Angel's Trumpet. They are *Datura arborea*, *D. cornigera* and *D. suaveolens*, and the last named has the additional attraction of fragrance. Both *D. cornigera* and *D. suaveolens* have double-flowered varieties and the double *D. cornigera* is sometimes sold as *D. knightii*, an erroneous name. Yet another kind with narrower more tubular flowers, yellow deepening to orange or red at the mouth, is *D. sanguinea*. It is hardier than the others and can be grown out of doors in some very sheltered places. All are quite happy outside from June to September and make handsome specimens for sunny sheltered patios and terraces. In winter keep them in a cool or intermediate house.

All can be grown in J.I.P. No. 2 compost in large pots or tubs. They should be watered freely in spring and summer and may be fed fortnightly in summer with weak liquid fertiliser. Plants can be pruned quite severely in autumn, so that they take up less room in the greenhouse, and will make well branched heads the following year. Increase is by cuttings of firm young shoots in spring in a propagator. Young plants may be restricted to a single stem until they are from 1 to 4 ft. high and then can be pinched to make them branch, so forming standards.

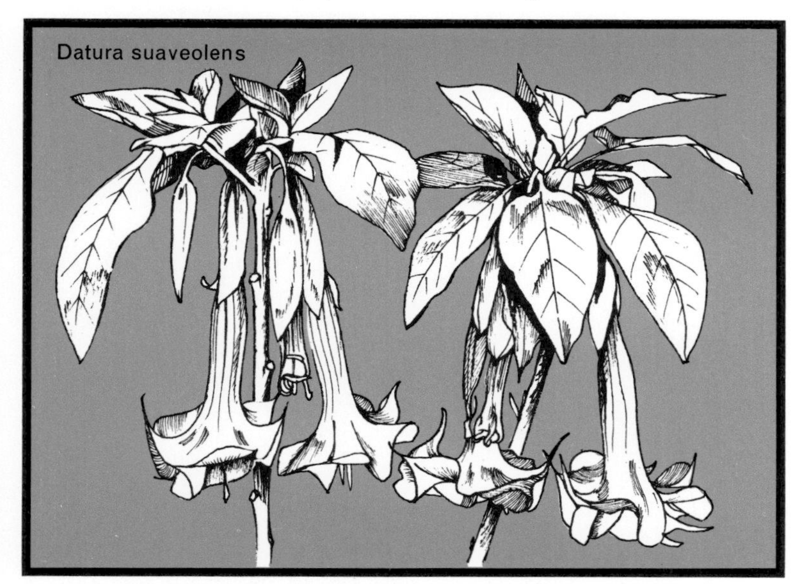

Didiscus and Others

The delicate clusters of flowers of *Didiscus caeruleus* are produced in summer and have earned it the common name of Blue Lace Flower. Didiscus (also called trachymene) is a half-hardy annual

Didiscus caeruleus, the Blue Lace Flower, has small lavender-blue flowers in flat clusters on slender 18-in. stems in summer and is easily grown in a cool house. Seed should be sown in spring in a temperature of about 15° C. (60° F.) in J.I.S. or peat seed compost, seedlings pricked out into the same compost and later potted singly into 4-in. pots in J.I.P. No. 1 or peat-based potting compost. They should be watered freely throughout, fed occasionally in summer and lightly shaded.

Trachelium caeruleum, the Blue Throatwort, also produces its small violet-blue or white flowers in clusters all summer on stems 18 to 36 in. high. Though a herbaceous perennial it is usually treated as an annual or biennial, being raised from seed sown in spring, or for longer, earlier flowering, in mid-summer. Spring-sown seedlings are grown in exactly the same way as didiscus except that 5-in. pots are likely to be required. Summer-sown seedlings are overwintered in a cold or cool house and may need to be potted on into 6- or 7-in. pots in spring.

The trailing varieties of the common lobelia and also *Lobelia tenuior* are usually grown as pot plants or in hanging baskets. Though perennial they are mostly grown as half-hardy annuals, seed being sown in early spring in a temperature of 15 to 18° C. (60 to 65° F.) and then grown on in the same way as didiscus, but without shade. In summer plants can be outdoors.

Epacris, Eucalyptus

Epacris are Australian plants which will flower in winter under cool greenhouse conditions. There are two species commonly grown, this one being *Epacris longiflora*

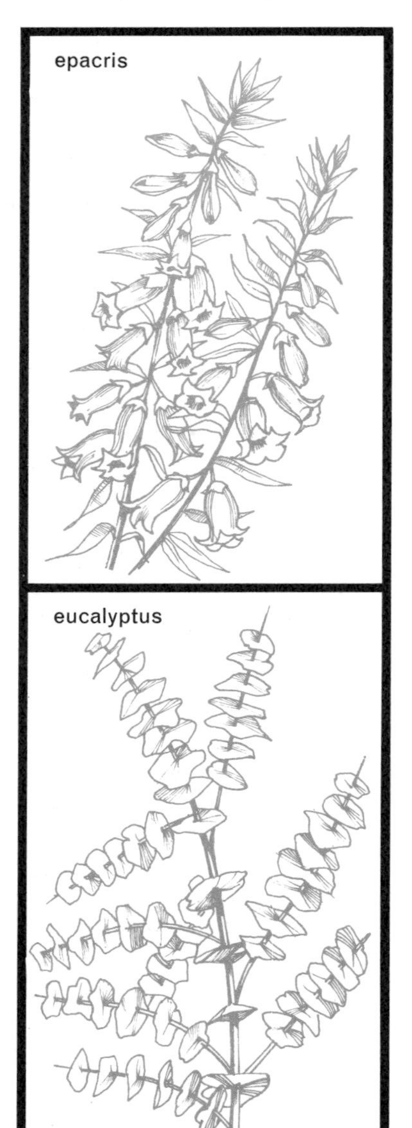

Epacris are small evergreen shrubs flowering in winter. They are known as Australian Heath and have a superficial resemblance to some of the South African winter-flowering heaths. They are, however, unrelated and have longer, more tubular flowers than most heaths. The colour range is from white to carmine. Plants are best grown in a peat-based compost in a cool greenhouse and can be accommodated in a frame in summer. No shading is required. They must be watered carefully throughout, the compost kept moist, not sodden. Plants should be pruned fairly severely each spring after flowering to keep them bushy. Increase is by seed sown on peat in a cool greenhouse or by cuttings of firm young shoots in a propagator.

Eucalyptus are the Australian evergreen Gum Trees. Most are on the borderline of hardiness, grow readily and rapidly from seed and stand hard pruning well. They can be grown as cool (or even cold) greenhouse pot plants, at least for a few years, after which they can be replaced with new seedlings. Seed germinates readily in spring in a temperature of 13 to 15° C. (55 to 60° F.). Seedlings should be potted singly in J.I.P. No. 1 or peat-based compost and be moved on as necessary into J.I.P. No. 2. They should be watered normally, fed with weak liquid fertiliser fortnightly in summer and grown without shade, and can be cut back each spring. In summer plants can stand out of doors.

Euphorbias

Showy, red bracts make the Poinsettia, *Euphorbia pulcherrima*, a very popular pot plant at Christmas time. Poinsettias are easy plants to cultivate in a heated greenhouse

This is a big family of plants, some succulents grown solely for their strange foliage and some hardy perennials. The three most important flowering kinds for the greenhouse are *Euphorbia pulcherrima*, *E. fulgens* and *E. splendens*.

Euphorbia pulcherrima is the Poinsettia, which is popular at Christmas time because of the handsome scarlet or pink bracts which surround the insignificant flowers. It is not a difficult plant to grow but needs the temperatures of an intermediate greenhouse. It is usually grown from cuttings of young shoots 2 to 3 in. long taken in spring and rooted in a propagator. When well rooted these are potted singly in 4-in. pots in J.I.P. No. 2 or peat-based potting compost. The temperature is gradually lowered and in summer the plants are grown in a lightly shaded house with a fairly humid atmosphere. Water freely and feed every 7 to 10 days with weak liquid fertiliser. Water moderately in autumn and winter and maintain a minimum temperature of 13° C. (55° F.). Plants in flower can be brought indoors for a few weeks but should then be returned to the greenhouse. After flowering, watering is reduced until for two or three weeks in early spring the soil is almost dry. Then stems are cut back to about a third, watering is resumed and the temperature is raised to at least 18° C. (65° F.) to start new growth and provide further cuttings.

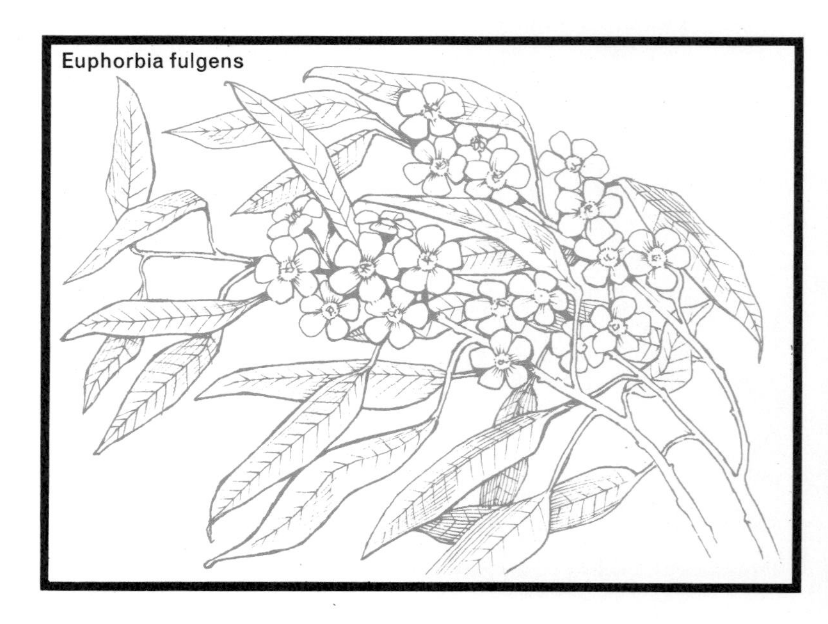

Francoa, Gerbera

The flowers of *Euphorbia fulgens* are, in fact, coloured bracts. This is also the case throughout the Euphorbia family, the true flowers being very insignificant

Euphorbia fulgens is a shrub with slender stems wreathed in winter with clusters of small orange-scarlet flowers. These stems are usually supported in such a way that they arch gracefully and display their flowers well. It is a plant requiring treatment very similar to that of *Euphorbia pulcherrima*, the Poinsettia, but it prefers a slightly higher temperature in winter and is happier in a warm rather than an intermediate house. The long flower stems are often cut for use in floral arrangements, and this does the plants no harm since it encourages branching from low down and, as with the Poinsettia, produces early new growth to be used as cuttings. However, since the plants are smaller and less leafy there is not the need to restart from cuttings each year.

Euphorbia splendens has orange-red flowers larger than those of *E. fulgens* and produced in summer. It is a stiffly branched shrub armed with spines and known as the 'Crown of Thorns'. It makes a good pot plant in J.I.P. No. 1 compost in a cool or intermediate greenhouse without shade at any time. Water freely in spring and summer, sparingly in autumn and winter. *E. splendens* grows well in a sunny window provided the temperature never drops below 7° C. (45° F.).

Gerberas are native to South Africa, hence their common name of Transvaal Daisy. These attractive daisy-like flowers are often used for flower arrangements and bouquets

gerbera

Francoas are known as Bridal Wreath because of their slender 2- to 3-ft. sprays of flowers in summer. *Francoa ramosa*, which is pure white, is the species most commonly seen, but *F. appendiculata*, in which the flowers are flushed with red, is also grown. Both are herbaceous perennials, hardy out of doors in very sheltered places and easily grown as pot plants in a cold or cool greenhouse or a well-lighted window. Grow in J.I.P. No. 2 or peat-based potting compost. Water normally and shade lightly in summer, at which season plants can be placed out of doors or in a frame if wished. Increase is by division in spring, the best season for re-potting.

Gerbera jamesonii is a beautiful South African plant known as the Transvaal or Barberton Daisy. The large daisy-like flowers produced from late spring throughout the summer have very narrow petals which may be pink, red, terra-cotta, orange or yellow. There are both single and double varieties. All are suitable for cultivation in a cool, well-lighted greenhouse. They need sharp drainage, careful watering and free ventilation except in cold weather. They are grown from seed sown in early spring or from cuttings in spring. Grow on in J.I.P. No. 1 compost with extra sand or grit, maintain a minimum winter temperature of 7° C. (45° F.) and do not shade at any time. Water sparingly in autumn and winter, fairly freely in spring and summer.

Fuchsias

Greenhouse fuchsias are accommodating plants for they can be grown as bushes, standards or in hanging baskets. These graceful flowers belong to the variety Coralle

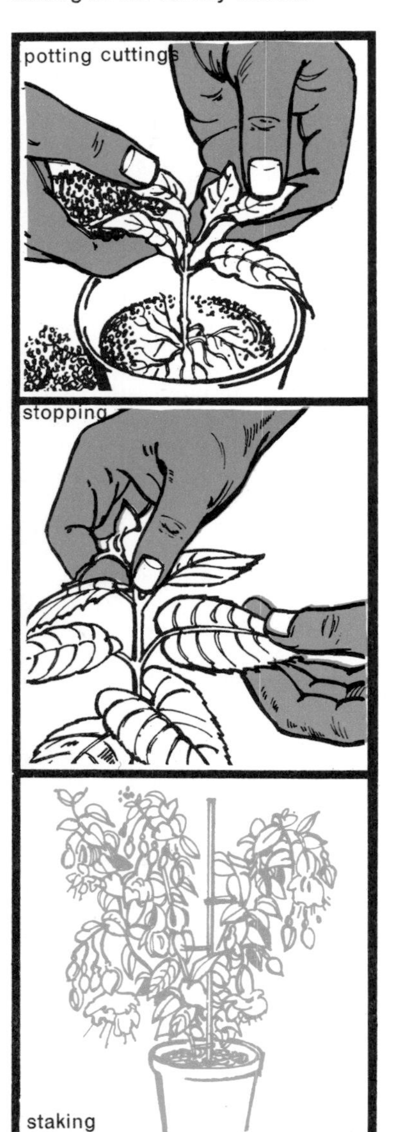

Fuchsias flower more or less continuously from May to October. There are a great many varieties; some have single flowers, some double, some are erect, some spreading or trailing, and all are easily grown from cuttings.

Cuttings of firm young shoots will root readily in a propagator at practically any time from spring to autumn but March–April and August–September are particularly convenient times. Pot cuttings as soon as they are well rooted into J.I.P. No. 1 or peat-based potting compost in 3-in. pots. Pot on into 5-in. pots and J.I.P. No. 2 or peat potting compost when the smaller pots are full of roots, and, if necessary, pot again into 6- or 7-in. pots.

Water freely in spring and summer and very sparingly in winter if the greenhouse is inadequately heated, but moderately if a minimum temperature of 10° C. (50° F.) can be maintained. No artificial heat is required from April to October. Shade from strong direct sunshine and ventilate freely. Most varieties can be placed out of doors from May to September.

Plants can be grown on one stem as standards, or if grown as bushes or for hanging baskets they should have the tip pinched out when 2 to 4 pairs of leaves have been made, and the sideshoots should be treated similarly. All sideshoots on the standards should be removed until the required height of the stem is reached, then treat as a bush

Globba, Ixora

Ixoras are evergreen shrubs of great beauty. They are natives of tropical areas and require warm greenhouse conditions for successful cultivation in this country

These are handsome tropical plants for well warmed greenhouses. Globbas are herbaceous in growth and belong to the ginger family. The flowers are curiously formed, carried in terminal sprays, red and yellow in *Globba atrosanguinea* and *G. schomburgkii*, magenta and yellow in *G. winitii*. Like many tropical plants they have a long and rather unpredictable flowering season.

Ixoras are evergreen shrubs, 2 to 3 ft. high with shining leaves and clusters of brilliantly coloured flowers, narrowly tubular in form except at the mouth which flares open like a star. Most of the varieties cultivated are garden raised hybrids and the colour range is from yellow to orange-red. All produce their flowers in summer.

Both plants should be grown in a warm greenhouse, the ixoras either in a peat-based compost or in J.I.P. No. 1 with extra peat, the globbas in normal J.I.P. No. 1 or peat-based compost. They should be watered freely in spring and summer, moderately in autumn and winter and can be fed fortnightly in summer with weak liquid fertiliser. During the summer they should be shaded from direct sunshine and a humid atmosphere should be maintained.

Ixoras can be increased by cuttings of firm young shoots in spring or summer in a propagator at a temperature of 21° C. (70° F.); globbas by division in spring, the best season for re-potting.

Gloxinias

Hybrid gloxinias, with their velvet-textured petals, are impressive plants to grow in the greenhouse or the home. The flowers can range in colour from white to pink, red and purple

Gloxinias are extremely showy plants with velvet-textured leaves and fine trumpet-shaped flowers in a range of brilliant colours. They are in flower from about mid-summer until early autumn and have tuberous roots. They can be grown either from these tubers or from seed.

Sow seed from January to March in John Innes Seed Compost or a peat-based seed compost in a temperature of 15 to 18°C. (60 to 65°F.), covering only with a piece of glass and paper laid over the pan. Remove the paper directly the seeds germinate and the glass two or three days later. Prick out the seedlings in similar compost while still small and when they have a few leaves, pot in 3-in. pots in John Innes No. 1 Potting Compost. As the pots fill with roots move the seedlings on to 4-in. and then to 5- or 6-in. pots, at the last stage using J.I.P. Compost No. 2.

Grow throughout in a temperature of around 15 to 18°C. (60 to 65°F.) shading from direct sunshine from May onwards, and water the seedlings freely, maintaining a fairly moist atmosphere by damping down at least once a day in summer. In October, gradually reduce the watering and from November keep the plants quite dry. Tubers can be shaken clear of soil and stored at not less than 10° C. (50° F.). In February or March set the tubers shoulder to shoulder in seedboxes of peat, water moderately and start at 15° C. (60° F.).

Gloxinia hybrida

Winter Heaths

Erica canaliculata is a good heather for cultivation under glass. Ideally, winter heaths need a special compost made up of 3 parts sphagnum peat and 1 part lime-free sand or grit

The winter heaths and heathers are all South African varieties of erica. Several of the best kinds, such as *Erica hyemalis* with pink and white flowers, *E. gracilis* with rosy-purple flowers and *E. nivalis* with white flowers, are grown in great quantities for the Christmas pot-plant trade. Unfortunately, commercial growers often trim the roots to such an extent before sending the plants to market that they have little chance of surviving more than a few weeks. It is better to start with young plants that have not yet flowered.

Pot these in a mixture of three parts sphagnum peat and one part lime-free sand or grit. Water carefully but adequately, keeping the soil moist and never allowing it to dry right out. Grow in a well-ventilated sunny greenhouse with a minimum winter temperature of 7° C. (45° F.). From June to September the plants will be better in a frame with free ventilation and with the pots plunged to their rims in moist peat to keep the compost from drying out. Shade only from the strongest sunshine.

Winter heaths can be increased in autumn by cuttings of firm side-shoots inserted in peat and sand in a temperature of 15° C. (60° F.). When rooted, pot in 3-in. pots and do not pot on until they are well filled with roots.

Erica hyemalis

Heliotrope, Humea

The fragrant flowers of heliotrope are produced in summer, and it is a popular bedding plant. Heliotropes do not need much heat at any time but are damaged when temperatures approach freezing point

Humea elegans

The common purple heliotrope or Cherry Pie, so popular as a summer bedding plant, also makes an excellent pot plant; and *Heliotropium peruvianum*, the wild plant from which the garden varieties have been developed, is sufficiently vigorous to cover a pillar or part of a wall if suitably supported.

Heliotrope can be grown from seed sown in spring in J.I.S. or peat-based seed compost and later potted singly in 4- or 5-in. pots in J.I.P. No. 2 compost. They should be watered fairly freely in spring and summer, moderately in autumn, sparingly in winter. Since they are so easily raised from seed many gardeners prefer to discard them after flowering.

Humea elegans is a biennial with slender 5-ft. stems bearing loose pendant sprays of small reddish-brown flowers in late summer and early autumn. It is known as the Incense Plant because of the distinctive aroma of its foliage. Seed should be sown in late spring or early autumn in J.I.S. or peat-based seed compost and germinated in a temperature of 15° C. (60° F.). Seedlings are potted singly in J.I.P. No. 2 or equivalent. Plants are over-wintered in a cool or intermediate greenhouse and moved on again in spring into 7- or 8-in. pots. Each stem must be staked and the plants given plenty of light. Great care must be taken not to overwater in winter and careful but adequate watering is essential throughout.

Jacaranda, Lagerstroemia

Jacarandas are grown in greenhouses and conservatories primarily for the beauty of their fine foliage; this is especially evident in young plants, as shown above

Lagerstroemia indica

Jacaranda acutifolia

In frost-free climates *Jacaranda acutifolia* (or *mimosifolia*, the name by which it used to be known) is one of the most admired of flowering trees. It can be grown just as readily in frost-proof greenhouses, the only difficulty being to find room to accommodate it. However, though young plants are unlikely to flower they are worth growing for their foliage. Plants can be raised from seed sown in J.I.S. or peat-based seed compost in a temperature of 15 to 18° C. (60 to 65° F.) in spring, seedlings being potted on as necessary in J.I.P. No. 1 or equivalent compost. Jacarandas should be watered normally and should be given as much sun and light as possible. They can be cut back in spring to delay the time at which they will get too large.

Lagerstroemia indica, the Crape Myrtle, is a deciduous shrub or small tree also much used for street planting in warmer climates. It can be grown readily in a frost-free greenhouse and, unlike the jacaranda, quite small specimens will produce the long terminal sprays of curiously crimped pink flowers. There are also white, mauve and heliotrope varieties. If desired it can be trained on the back wall of a lean-to greenhouse or conservatory. Treatment is as for the jacaranda except that plants are raised from cuttings of firm young shoots in spring or early summer in a propagator. Fairly hard pruning each spring not only keeps plants in bounds but encourages flower production.

Lantana, Verbena Lippia and Others

Lantana camara is a colourful shrub which has a long flowering season and can be used for bedding out in summer. It must be pruned hard back each spring or it will become straggly

Nerium, popularly known as oleander, makes a good greenhouse plant and under the right conditions it will reach a height of 6 ft. or more. The leaves, stems and flowers are poisonous

These plants are related and have a strong resemblance but the lantanas are, in general, more bushy than the verbenas. Both are readily grown in any sunny, frost-proof greenhouse.

The garden varieties of lantana are usually 2 or 3 ft. high and have pink, red, yellow, orange or white flowers produced continuously from spring to autumn. Seed can be sown in spring in J.I.S. or peat-based seed compost in a temperature of 15 to 18° C. (60 to 65° F.), seedlings being potted in J.I.P. No. 1 or equivalent compost and grown on in a cool or intermediate greenhouse without shading. They should be watered normally and can be pruned fairly severely each spring to keep them bushy and compact. Alternatively, cuttings of firm young shoots can be rooted in a propagator.

There are many different verbenas but the best to grow as pot plants are those marketed as half-hardy annuals for summer bedding. In fact, though readily raised from seed in exactly the same way as lantanas, they are not annuals but half-hardy perennials and they can just as easily be increased by cuttings. They are sprawling or bushy plants with showy flowers in a wide range of colours including pink, red, crimson, blue, mauve, and white. Culture is the same as for lantana and the flowering season almost as long. Many people will prefer not to attempt to overwinter them but to renew annually from seed.

All these are shrubs which can be grown in a frost-proof greenhouse in large pots or tubs, or planted in beds of soil. Lippia is grown for its strongly lemon-scented leaves and is popularly known as lemon-scented verbena. Luculia has showy heads of sweetly scented pink flowers in late summer and autumn. Nerium is popularly known as oleander and has showy rose or white flowers. Its leaves, stems and flowers are poisonous.

Grow all three in J.I.P. No. 1 compost or in good loamy soil with a scattering of bonemeal and re-pot when necessary in March or April. Grow in full sunlight without shading throughout and water freely in spring and summer, very sparingly in winter, though luculia must be kept well watered through the period it is in flower, which may well be into November. Keep luculia in the smallest pots or tubs that will contain the roots or, if planted in a border, contain the roots with bricks or slates to promote free flowering.

Shorten the stems of luculia and nerium in December or January; cut back lippia almost to the base in February. Maintain a minimum winter temperature of 7° C. (45° F.) for luculia and nerium and of 2° C. (35° F.) for lippia. No artificial heat is required for any of these plants from April to October. Increase by summer cuttings.

Medinilla, Oxalis

The fine flowers of *Medinilla magnifica* make this one of the most eye-catching of plants. It comes from tropical areas and must be given warm conditions with a minimum winter temperature of 15°C. (60°F)

Oxalis purpurata bowiei

Medinilla magnifica really is a magnificent plant with large, shining evergreen leaves and fine hanging trusses of rosy-pink flowers topped by lighter pink bracts in late spring and early summer. It always excites admiration but it does need a warm greenhouse with a minimum winter temperature of 15° C. (60° F.). It can be grown in J.I.P. No. 2 compost with extra peat to provide a rather spongy mixture. Water freely in summer when a moist atmosphere must be maintained with a fair amount of shade. In winter only sufficient water should be given to keep the soil just moist, with increasing supplies as growth restarts in spring. Plants can be fed every fortnight from late spring throughout the summer. Increase is by cuttings of firm young growth in spring in a propagator at 20 to 24° C. (70 to 75° F.).

There are numerous species of oxalis suitable for greenhouse cultivation, all small tuberous-rooted plants with somewhat clover-like leaves and loose sprays of flowers produced mainly in spring. *Oxalis deppei*, rose, *O. purpurata*, purple and its larger variety *bowiei* are recommended. All can be grown in J.I.P. No. 1 or peat-based potting compost, about 5 little tubers in each 4- or 5-in. pot, in a cool greenhouse without shade. Pot in autumn, water sparingly in winter, freely as growth increases, then after flowering gradually reduce the water supply and keep quite dry in late summer until it is time to re-pot and start the tubers.

Mimulus, Reinwardtia

Reinwardtia trigyna is an evergreen shrub which produces its bright, showy flowers at a most useful period from autumn to spring, when colour can be rather scarce

Mimulus aurantiacus

The best mimulus for the greenhouse is *Mimulus aurantiacus*, sometimes still called by its old name of *Diplacus glutinosus*. It is a small, rather straggly evergreen shrub which produces a non-stop display of orange or coppery-red flowers from spring to autumn. It will grow 3 ft. high but can be kept shorter and more bushy by spring pruning. It should be grown in J.I.P. No. 2 or peat-based compost in a cold or cool greenhouse with all the light and sunshine possible. If necessary it can stand out of doors throughout the summer. Water fairly freely in spring and summer, very sparingly in autumn and winter.

The hybrid Monkey Flowers or mimulus usually sold as half-hardy annuals for summer bedding also make showy cold or cool greenhouse pot plants. They can be grown in J.I.P. No. 1 or peat-based potting compost, should be watered freely and shaded from direct sunshine. They are best discarded after flowering.

Reinwardtia trigyna is an evergreen shrub about 2 ft. high with showy yellow flowers produced from autumn until spring. Though it will grow in a cool house, flowering is more reliable and continuous in an intermediate house. Grow in J.I.P. No. 2 compost or plant in a bed of good loamy soil. Water normally and feed occasionally in summer with weak liquid fertiliser. Do not shade at any time. Prune fairly severely each spring as soon as flowering stops.

Myrtle, Leptospermum

Red Damask is a free-flowering variety of *Leptospermum scoparium*. Leptospermums are unusual plants for the greenhouse and, as most kinds are vigorous, they are better grown in borders of soil

The common myrtle, *Myrtus communis*, is a shrub with evergreen leaves, small, white sweetly scented flowers freely produced in summer, followed by blue-black fruits that are edible. It can be grown outdoors in sheltered places, but it is safer in a cool house and its compact variety makes an excellent pot plant. As an alternative to a tub which can be wheeled out in summer it can be planted in a bed of good loam, and if of the ordinary variety, its stems can be trained to trellis or wires. In pots or tubs plant in J.I.P. No. 2 compost, water normally and in the house give sufficient shade in summer to break direct sunshine. There are several other kinds such as *M. luma* with handsome cinnamon and cream bark; *M. lechnreiana* which flowers (on very young plants) in May; and *M. ugni*, which has the most palatable fruits. All can be grown in the same way but preferably in beds of soil.

Leptospermum scoparium is a New Zealand evergreen shrub which can be grown in exactly the same way as the myrtles. The leaves are very narrow, the flowers small, white, pink or carmine, single or double and very freely produced in May and June. For pot cultivation the best variety is *nanum* which only grows about 1 ft. high against 6 ft. or more for most of the others.

Both myrtles and leptospermum can be increased by summer cuttings in a propagator.

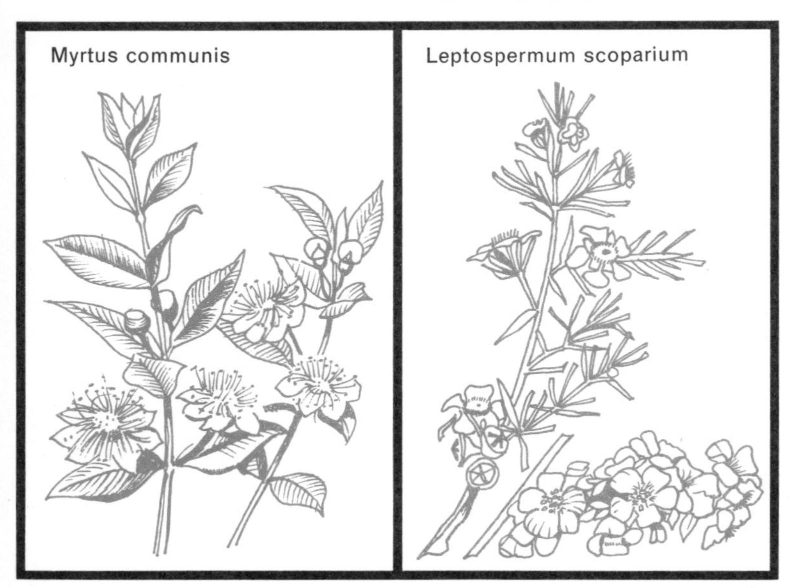

Nepenthes, Sarracenia

Few plants are more bizarre in their appearances than those which are insectivorous. *Sarracenia purpurea* is just such a fascinating plant, though it is unnecessary to provide it with insects

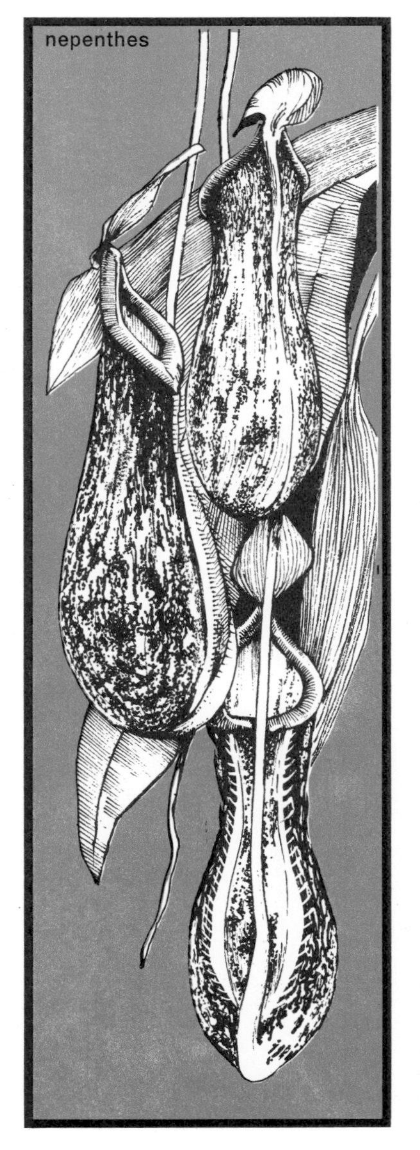

Both nepenthes and sarracenias are known as Pitcher Plants because they catch and digest insects in pitcher-like structures. Those of nepenthes being rather like old-fashioned pipes hanging on slender stems and in some species highly ornamental; those of sarracenia are narrower and more vase like and grow up from the ground. In cultivation it is unnecessary to give them insects.

Nepenthes are usually grown in hanging baskets, often made of teak like those used for some orchids. The compost used is 2 parts peat and 1 part chopped sphagnum moss and plants are re-potted when necessary in March. They require warm house conditions with a very humid atmosphere at all times and shade from all direct sunshine throughout the year. Water freely from March to September and moderately from October to February. Increase is by cuttings of one-year-old shoots in peat and moss in a propagator at 26 to 30° C. (80 to 85° F.).

Sarracenias are much hardier and can be grown in a cool greenhouse, but they also need a damp atmosphere and shade and the same mixture of peat and chopped sphagnum moss. They can be grown in pots and are very suitable for Wardian cases or plant cabinets. Water freely from April to October, sparingly from November to March and syringe daily with water in summer. Increase is by division in March or April which is also the best time for re-potting.

Pelargoniums

Regal pelargoniums provide a brilliant show of colour in May, June and July. The many varieties cover a very wide colour range. This one is called Stardust

Pelargonium is the correct name for the greenhouse and summer bedding plants commonly called geraniums. There are numerous types, classified as Zonal-leaved, Ivy-leaved, Scented-leaved and Regal, and all can be grown in a greenhouse with a minimum temperature of 10° C. (50° F.).

All can be grown from cuttings of firm young shoots taken in spring or late summer. When well rooted, pot the cuttings singly in 3-in. pots in J.I.P. No. 1 or peat-based compost and grow on in a light greenhouse. Water moderately and maintain a fairly dry, airy atmosphere. Pot on into 5-in. pots and J.I.P. No. 2 or peat potting compost when the smaller pots are full of roots. The old plants should be fed fortnightly in summer.

Place ivy-leaved pelargoniums at the edge of the staging or in hanging baskets so that they can trail naturally, or train them up trellis or canes. Pick off the flower buds of scented-leaved pelargoniums as the flowers are small and unattractive. Cut back regal pelargoniums after they have flowered in June or July and stand them out of doors, making cuttings from the new shoots when they appear.

Zonal-leaved and ivy-leaved pelargoniums will flower all summer and even in autumn if some flower buds are picked off to rest the plants, but those bedded out must be brought back into the greenhouse in late September.

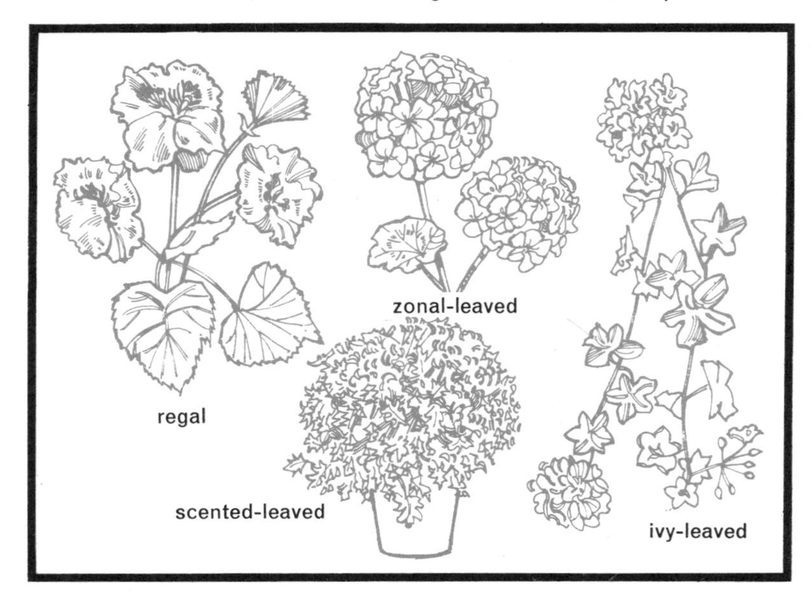

Petunia and Others

Petunias, which might well be considered the most popular of all bedding plants. The colour range includes many shades of red, purple, blue, salmon, pink, yellow and white

Petunias are grown from seed sown in spring in J.I.S. or peat-based seed compost in a temperature of 15 to 18° C. (60 to 65° F.). The seedlings are pricked out in the same compost and later potted singly in 4-in. pots in J.I.P. No. 1 or peat-based potting compost. No artificial heat should be necessary after April. Grow in a sunny greenhouse without any shade and water fairly freely throughout, but be careful not to splash the water over the flowers. Feed once a week from June onwards.

Impatiens holstii and *I. sultani* are similar plants popularly known as Busy Lizzie. They have gay pink, red, purple or white flowers which with sufficient warmth can be produced throughout the year. They are perennials often grown from seed in the same way as petunias, but with shade from strong sunshine in summer. Plants can also be grown from cuttings of firm young shoots in spring or summer. In a cool or intermediate house or a sunny room plants can be over-wintered easily and kept for years. Water moderately in autumn and winter. The variety with variegated leaves must be increased by cuttings.

Vinca rosea, the Madagascar Periwinkle (*now Catharanthus roseus*), resembles the Busy Lizzie and like it can flower all the year. Treatment is the same as for impatiens but it needs an intermediate or warm greenhouse and prefers J.I.P. No. 2.

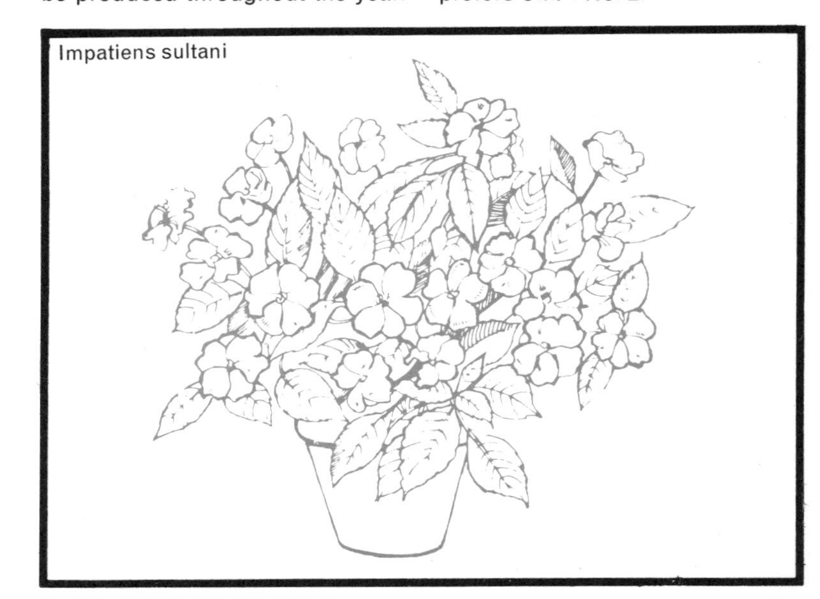

Primulas

The yellow flowers of *Primula kewensis* make it quite distinct from other greenhouse primulas as this colour is missing among the other groups. It is pleasantly scented

Of the four primulas which will flower in a greenhouse during the winter months, *Primula obconica*, *P malacoides*, *P. sinensis* and *P. kewensis*, the first two are the most rewarding. *Primula obconica* carries its quite large, blue, pink, salmon, crimson or white flowers in loose heads, and will keep on flowering most of the year. *P malacoides* has smaller flowers in larger, loose sprays, the colour range being pink and heliotrope to light crimson. *P. sinensis* has flatter heads of large pink, salmon, orange, crimson or blue flowers with fringed or waved petals. *P. kewensis* has sprays of scented yellow flowers and the leaves of some varieties are dusted with a mealy powder, or farina.

Sow seeds of *Primula obconica* and *P. sinensis* in March or April and *P. malacoides* and *P. kewensis* in May or June in J.I.S. or peat-based seed compost and germinate in a temperature of 15°C. (60°F.). *P. obconica* and *P. sinensis* germinate best in the light, so do not cover the seed pans with paper. Prick out the seedlings into J.I.P. No. 1 or a peat-based potting compost and, when they touch in the boxes, pot singly in 3-in. pots in similar compost. Pot on into 5-in. pots and J.I.P. No. 2 or peat-based potting compost.

Water fairly freely and grow in a cool, well-ventilated greenhouse or, from June to September, in a frame. From October on keep at a minimum of 7°C. (45°F.).

Protea and Relatives

Protea reflexa is one of the more unusual proteas. These South African plants have handsome flowers which can be used to make highly effective flower arrangements

The proteas are handsome shrubs with leathery leaves and extraordinary egg-shaped or conical flower heads, often 6 in. across, the tightly packed flowers encased in overlapping pink, red or yellow bracts. Allied to them and with flower heads of similar form are leucadendron and leucospermum.

All can be grown in cool or intermediate greenhouses or conservatories in large pots or tubs in peat-based potting compost or a mixture of 3 parts peat or leafmould, 2 parts sharp sand and 1 part lime-free loam. Alternatively, they may be planted in a bed of peat with which some lime-free loam and coarse sand has been mixed. A fairly open, free draining soil is essential. They also need plenty of sunlight and should not be shaded at any time of the year. If in pots or tubs they can be stood out of doors from June to September in a sunny, sheltered place. Water fairly freely from March to September, very sparingly from October to February.

Proteas and their relatives can be raised from seed sown in the autumn or late winter in the soil mixture recommended above and germinated in a temperature around 15° C. (60° F.) without any shading. Seeds germinate better if well soaked in warm water before sowing. Prick out singly into 3- or 4-in. pots in the same compost and pot on before roots penetrate through the drainage holes as proteas resent root breakage.

Rehmannia and Others

Ruellia macrantha is a semi-shrubby plant which grows 2 to 3 ft. high and makes a decorative pot plant provided it is cut back hard each spring after flowering

Rehmannia angulata is a showy, easily grown plant. Its 2- to 3-ft. flower stems are freely produced in summer and carry almost throughout their length light purple bell-shaped flowers. Though a herbaceous perennial it is often grown as a biennial from seed sown in early or mid-summer in an unheated greenhouse or frame. Seedlings are grown on first in J.I.P. No. 1 and later, when they reach the 6-in. pots in which they will flower, in J.I.P. No. 2 compost. Water normally throughout and grow in a frost-proof house shaded in summer. Pinch out the tips of plants when about 9 in. high.

Ruellia macrantha is semi-shrubby, 2 to 3 ft. high with heads of rosy-purple trumpet-shaped flowers in winter and spring. It can be grown in a cool house but probably will not flower so early or reliably as it will in an intermediate house. It should be grown in J.I.P. No. 2 with normal watering and shade from direct sunshine. Prune hard each spring after flowering and increase by cuttings of firm young shoots in a propagator.

Torenia fournieri is a half-hardy annual, 1 ft. high, producing in summer and autumn trumpet-shaped violet-purple and blue flowers with a yellow throat. It should be raised from seed sown in spring in J.I.S. or peat-based compost. Pot singly in J.I.P. No. 1 or equivalent and grow on in a light, frost-proof greenhouse with normal watering. Give some support to the slender stems.

Roses

Chicago Peace, an attractive sport of the famous rose, Peace, is a good variety for greenhouse cultivation in pots. The flowers are particularly large and slightly fragrant

Any varieties of rose may be grown in greenhouses and this is the best way to enjoy some of the rather tender old kinds such as the climbers Niphetos, white, and Maréchal Niel, sulphur yellow. But these take up a good deal of room and are most suitable for training up the back walls of lean-to greenhouses or conservatories. They should be planted permanently in a bed of rich loamy soil and treated in much the same way as outdoor roses. No artificial heat is needed, or only enough to exclude frost.

Bush varieties are more suitable as pot plants and those with very shapely flowers are most popular. Young plants should be potted in 8- or 9-in. pots in autumn in J.I.P. No. 3 or equivalent compost. They can stand outdoors or in a frame until mid-winter when they should be pruned really hard and left for a further three or four weeks. They can then be brought into a light greenhouse, either unheated or slightly heated according to how early flowers are required. Subsequently they are watered freely and, when flower buds appear, fed every 10 days with weak liquid fertiliser. They should also be sprayed occasionally with a greenfly killer. After flowering they are moved out of doors and pots can be plunged to their rims. They must be well cared for all summer and in autumn should either be repotted or generously topdressed with rich soil, some of the old topsoil being removed to make room.

Saintpaulias

Saintpaulias are commonly known as African Violets. There are varieties with violet, purple, red, pink or white, single or double flowers, which, in the right conditions, are produced almost continuously

These are the plants commonly known as African Violets. They make almost flat rosettes of velvety dark green leaves and produce violet, purple, pink or white, single or double flowers almost continuously throughout the year.

Grow saintpaulias in J. I. P. No. 1 with 25 per cent. extra peat or in a peat-based potting compost. Good plants can be obtained in 3½-in. pots. They should be kept in a minimum temperature of about 13° C. (55° F.) away from direct sunshine and in as moist an atmosphere as possible. Saintpaulias do well in plant cabinets, Wardian cases and bottle gardens. Water freely in spring and summer and moderately in autumn and winter when it is particularly important to avoid splashing the leaves or allowing water to lodge in the crowns of plants.

Increase saintpaulias by sowing seed in spring or summer in John Innes or peat seed compost and germinate in a temperature of 18° C. (65° F.). Prick off seedlings into similar compost and when their leaves touch, pot them singly in 3- or 3½-in. pots. Alternatively, grow saintpaulias from well-developed leaves removed in summer complete with leaf stalk and inserted like cuttings, leaf stalk first, in peat and sand in a propagator with a temperature of 18° C. (65° F.).

Smithiantha, Gesneria

Smithianthas are very attractive greenhouse plants producing spikes of tubular yellow, pink, orange or scarlet flowers in summer, well complemented by velvety leaves

These plants are closely related and their correct nomenclature is somewhat confused. All have tubers which are stored dry in winter in the soil in which they have been growing. Shake out these tubers in March and re-pot them singly in 5-in. pots in peat-based compost or a mixture of equal parts loam, leafmould and peat with 2 parts of sand. Only just cover the tubers, water rather sparingly at first, then freely and keep in a temperature of about 13° C. (55° F.) rising to 15 to 21° C. (60 to 70° F.) in summer. Shade from direct sunshine. After flowering gradually reduce the water and store pots on their sides quite dry from November to March in a frost-proof greenhouse.

They can also be grown from seed sown in January or June in a temperature of 15 to 18° C. (60 to 65° F.). Seedlings from a June sowing are kept growing all the first winter in an intermediate house and will start flowering the following spring. Seedlings from a January sowing will start to flower in late summer and are rested in winter like plants grown from tubers. The seedlings are raised in peat seed compost, pricked out into the same compost, and later potted individually like the tubers.

All these plants have velvety leaves and are 12 to 24 in. high. Smithianthas, sometimes called naegelia and popularly known as Temple Bells, are in many colours, e.g. yellow, orange, pink and red. *Gesneria cardinalis* is orange red.

Solanum, Ardisia

Solanum capsicastrum is often known as the Winter Cherry because of the brightly coloured berries which are produced during the late autumn and winter. Attractive bushes can be grown in 5- or 6-in. pots

The Winter Cherry, *Solanum capsicastrum*, is a very popular winter house plant because of its abundant orange fruits the size of cherries. Purchase plants then, or grow them from seed sown in a warm greenhouse or a sunny window in February or March. Pot seedlings in John Innes No. 1 or peat potting compost and stand them out of doors in a sheltered, sunny place from June to September, pinching out the tips of shoots occasionally to make the plants bushy. Syringe daily with water while the plants are in flower in early summer, water freely from April to October, moderately at other times and feed from May to August. During the winter keep in as light a place as possible, in a minimum of 7° C. (45° F.).

Ardisia crispa, also sometimes known as *A. crenulata*, is grown for its scarlet berries and these remain decorative from autumn until spring. It is a neat evergreen shrub which makes an excellent pot plant for a cool or intermediate house. Grow in J.I.P. No. 2 or equivalent compost, water normally, shade lightly in summer and ventilate as freely as outside temperatures allow while plants are in flower to assist pollination. Syringe frequently in summer. Overgrown plants can be pruned in late winter before sap starts to rise strongly. Increase is by seeds sown in spring in a temperature of 18° C. (65° F.), or by cuttings of firm young shoots in a propagator.

Sparmannia, Prostanthera

The evergreen shrub, *Sparmannia africana*, has large leaves and is an ideal plant for the bigger greenhouse as it can reach a height of 10 ft. unless pruned periodically

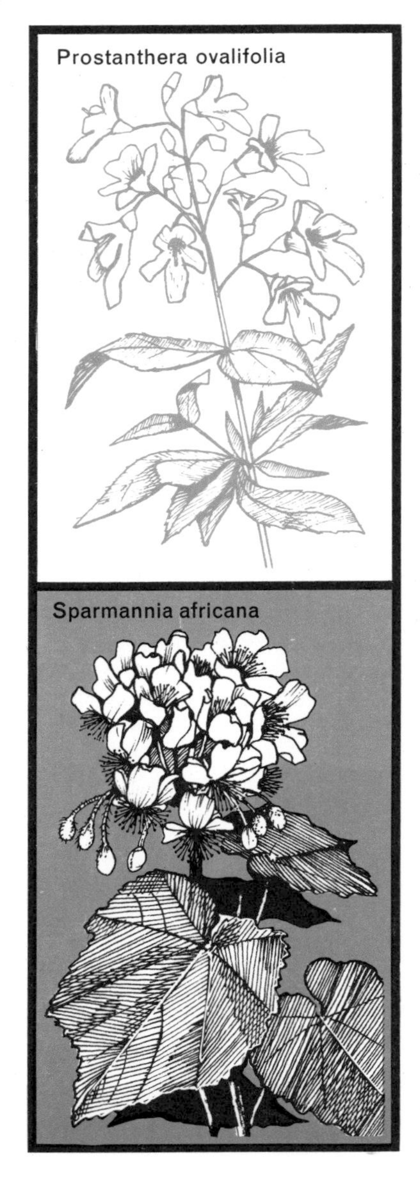

Sparmannia africana is a big, bushy evergreen shrub with large, soft, green leaves and white flowers with purple stamens, produced almost throughout the year. It will grow to 10 ft. but can be kept down to 3 or 4 ft. by periodic pruning, but this inevitably interferes with the continuity of flowering. Cultivation is similar to that for pelargoniums, plants being grown in J.I.P. No. 2 or peat-based potting compost in a light, cool or intermediate house. They should be watered normally and can be stood outdoors in a sunny place in summer. Increase is by cuttings of firm young shoots in spring or summer. Late winter is the best time for any really hard pruning.

The prostantheras are also evergreen shrubs suitable for cultivation in a cool greenhouse, either trained against a wall or grown in pots and pruned each year after flowering to keep them from becoming straggly. Three of the most popular kinds, *Prostanthera ovalifolia*, *P. rotundifolia* and *P. violacea*, have small leaves, clusters of small lilac or purple flowers in spring and will reach a height of 6 to 8 ft. The leaves are aromatic, hence the common name of Mint Bush. All three are sufficiently hardy to be grown outside in some very sheltered places, and, under glass, only require frost protection. Water normally and do not shade. Pot-grown plants can stand outdoors in summer. Grow in J.I.P. No. 1 compost and increase by cuttings of firm young growths in summer.

Strelitzia and Others

The exotic flowers of strelitzia appear in spring and early summer and it is easy to see how the common name, Bird of Paradise Flower, has arisen. Strelitzias thrive in a sunny position

Musa is the name of the fruit producing banana, which for ornament is grown primarily for its very large leaves, though greenhouse plants will also produce crops. Most kinds are too large to make good pot plants and are best grown in tubs or planted in a border of good soil, but *Musa sanguinea* is 4 ft. and can be grown in a large pot. *M. basjoo* is the hardiest and can be grown in any frost-proof greenhouse. Culture is as for strelitzia but with rather more water in winter and a very moist atmosphere in summer.

Heliconias have broad banana-like leaves and stout stiffly erect flower spikes bearing brightly coloured boat-shaped bracts. In some varieties the flower stem is red and in some the leaves are marked or veined with red. Heliconias should be grown in a warm greenhouse as for bananas, but with little water in winter. Shade well in summer. All these plants belong to the banana family. *Strelitzia reginae* is known as the Bird of Paradise Flower because its orange and blue flowers look rather like the head of a crested bird. It has broad leaves, reaches a height of 3 ft. and flowers in spring and early summer. It should be grown in large pots in J.I.P. No. 2 or can be planted in a border of good soil with some peat and rotted manure. It needs a cool or intermediate house, should be watered freely in spring and summer, sparingly in autumn and winter. Re-pot in spring and divide old plants.

Streptocarpus, Streptosolen

Streptocarpus are excellent subjects for the cool greenhouse. Their funnel-shaped flowers are available in purple, blue, red and white and they are in bloom from spring through to autumn

Streptocarpus have funnel-shaped blue, pink, red or white flowers on 9- to 12-in. stems from late spring to autumn. They are usually grown from seed though they can also be increased by leaf cuttings in summer in a propagator. Sow the seed in J.I.S. or peat-based seed compost in February or March and germinate in a temperature of 15° C. (60° F.). Prick out the seedlings in similar compost and later transfer singly to 3-in. pots in J.I.P. No. 1 or peat-based potting compost. Water fairly freely throughout and shade from direct sunshine. No artificial heat is required from May to September, but in autumn a minimum of 13° C. (55° F.) should be maintained, which can drop to 7° C. (45° F.) after flowering.

Streptosolen jamesonii is a sprawling evergreen shrub which can readily be grown as a climber to a height of 6 to 7 ft., or be trained over a wire balloon frame. Alternatively, if the tips of the shoots are pinched out occasionally it can be made into a more compact bush. It has clusters of bright orange flowers in summer and is a showy plant easily grown in any frost-proof greenhouse. Grow it in J.I.P. No. 2 compost, water normally and do not shade at any time. Shorten the previous year's stems fairly severely each February. Increase by cuttings of firm young shoots in spring or summer in a propagator.

Greenhouse Bulbs

Strictly speaking bulbs are modified buds built up of layers of overlapping fleshy scales, as can be seen very clearly when an onion is cut in half and the scales fall apart. In some bulbs, particularly lilies, the scales are so loosely arranged that they are quite obvious, in others, such as the hyacinth, they are so closely packed that outwardly the bulb appears solid.

As a rule when gardeners speak of bulbs they are extending the term to include corms, which really are solid right through, except for the membranous coat on the outside. Typical examples familiar to most people are the corms of gladiolus and crocus, and if these are cut in half their solid structure will become apparent.

This difference is really a technicality of interest only to scientists. Because of its affinity to a bud a bulb is often a complete plant in embryo. Packed tightly within it are leaves and flowers, and since the outer scales themselves are packed with food, moisture and warmth are all that are required to bring the plant to maturity. This can be seen most dramatically with hyacinths, which are often grown in special glasses designed to support them just above water into which their roots can descend. Placed in a sunny window and given no attention other than maintaining the level of water in the glass, the hyacinth will grow strongly and produce flowers of excellent quality.

By contrast, a corm is purely a storage organ, really a thickened stem with no leaves or flowers coiled within it ready to develop if given the correct conditions, though these are in the buds beneath the scales and care must be taken not to get them knocked off. The corm exists to give the plant a good start when conditions for growth are right, but it packs no guarantee within itself that the growth will culminate in flowers.

The moral of this is that good fat bulbs provide about the most foolproof method of starting to garden. It really is a very satisfying experience for a beginner to be given, say, a fine hippeastrum bulb and a few simple instructions and to be rewarded in a matter of weeks with a display of opulent flowers sufficiently good to win a prize in any show.

But to be sure of this happy result the bulbs must be good. Small bulbs will contain small flowers or no flowers at all, though if grown on well they may produce flowers another year. For instant success, however, one must buy good bulbs and these naturally enough cost more than small bulbs of the same variety.

This emphasis on size is nothing like so important with corms. Very small ones may only be suitable for growing on and building up stock for future years, but medium size corms can be just as satisfactory (occasionally even more satisfactory) than the largest, which may be more expensive. So when shopping for bulbs and corms spend your money where it will produce the best results.

Acidanthera, Watsonia

The handsome watsonias are not difficult plants to grow provided a frost-proof greenhouse or some other form of protection is available during the winter

Acidantheras look a little like gladioli, producing fragrant, white, maroon-blotched flowers on 2-ft. stems in August and September. They can be grown out of doors in well-drained soils and warm, sheltered places, though it is best to lift the corms in the autumn and store them in a frost-proof place until the spring. They al o make excellent pot plants for an unheated greenhouse.

Pot in spring four or five corms in each 4-in. pot in John Innes No. 1 or peat-based potting compost. Water moderately at first, then freely as growth starts, but reduce the water supply after flowering and keep the corms quite dry from November until February.

Watsonias are also allied to gladioli but the flowers are more tubular and the sword-like leaves are evergreen. In mild places, particularly near the sea, they can be grown out of doors but in most parts of the country they must be kept in a frost-proof greenhouse at least during the winter, or be covered with cloches or frames. Apart from this they are not difficult to grow. They have no resting season, so water moderately throughout the year, increasing the amount slightly as the flower spikes form.

There are a great many different kinds of watsonia, of which *W. ardernei*, white, *W. beatricis*, apricot to orange-red and *W. stanfordiae*, rose purple or scarlet, are typical.

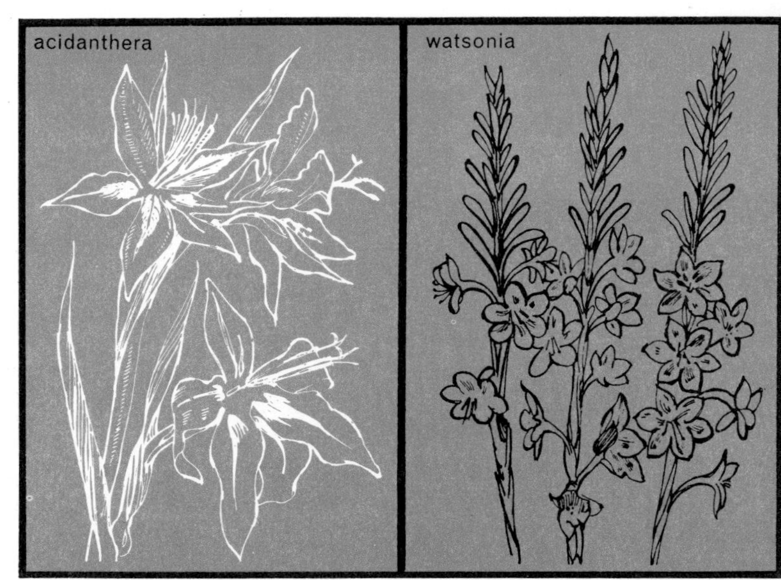

Agapanthus, Clivia

Clivia nobilis, with its large clusters of trumpet-shaped flowers in late winter and early spring, is an impressive plant for the home or greenhouse

The African lilies or agapanthus, which are not true lilies, are, like clivia, natives of South Africa. They both make fleshy roots and have rounded heads of flowers, blue or white in the African lily, orange-red or yellow in clivia.

The big African lily, *Agapanthus umbellatus,* is often grown in large pots or tubs so that it can be stood out of doors in summer when in flower and removed to a frost-proof greenhouse in winter. The small African lilies, *A. mooreanus* and Headbourne Hybrids, are hardier and can be grown out of doors in sunny sheltered places.

Clivias are also grown in large pots or tubs, but they flower in late winter and spring, too early to be put out of doors. Like *Agapanthus umbellatus* they require complete protection from frost in winter.

Grow these plants in good, loamy soil or in John Innes No. 2 Potting Compost, watering freely in spring and summer, and sparingly in autumn and winter. Do not shade the plants at any time of the year.

Increase them by dividing the roots in spring, or after flowering in the case of clivia.

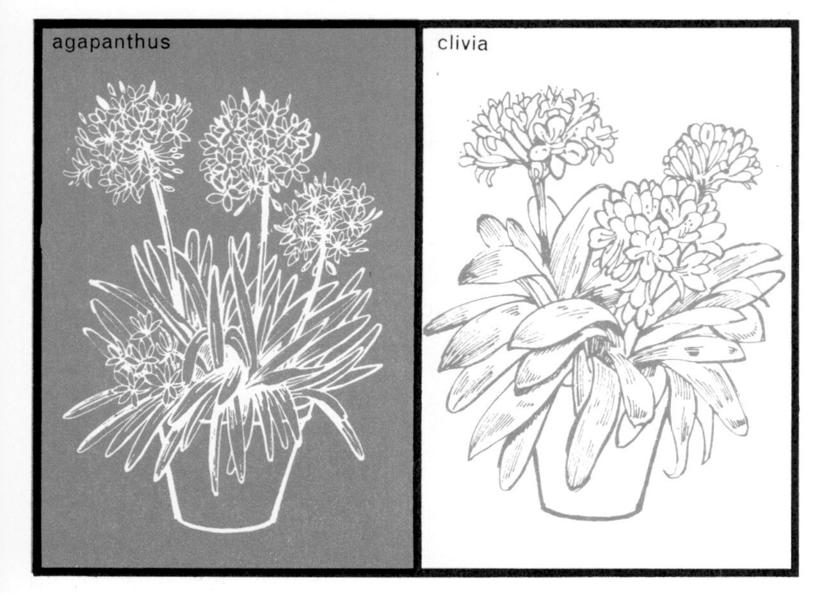

Babiana and Others

Babianas are excellent subjects for the cool greenhouse and in some favoured districts they may be grown out of doors. The trumpet-shaped flowers are in a range of bright colours

Babianas, ixias, sparaxis, sprekelias and tigridias can all be grown in a sunny, frost-proof greenhouse. Babianas have little sprays of blue, rose or carmine flowers on 6- to 12-in. stems in May and June. Ixias have sprays of starry red, crimson, orange, yellow or white flowers on wand-like 2-ft. stems in spring. Sparaxis, kncwn as the Harlequin Flower because it often combines two contrasted colours, may be red, crimson, purple, yellow, orange or white, is 12 to 15 in. high and spring flowering. Sprekelia, or Jacobean Lily, has large scarlet, spidery-looking flowers in June on stout 1-ft. stems. Tigridias, Tiger Flowers, also 1 ft. high, have gleaming red, pink, orange or yellow flowers, often handsomely spotted with one colour on another. Each bloom only lasts for a day, but a succession of buds opens for weeks in summer.

Grow all these bulbs in J.I.P. No. 1, placing 5 to 7 bulbs in each 4- or 5-in. pot, except for sprekelia where one bulb to each pot is sufficient. Plant babianas, ixias and sparaxis in autumn and over-winter in a cool house. Plant sprekelias and tigridias in spring in a cold or cool house. Water all sparingly at first, freely while they are growing, but gradually reduce the water supply after flowering and then keep the bulbs dry until re-potting. Over-winter those for spring planting in a frost-proof cupboard or room.

All can also be raised from seed sown in a cool house in spring.

Eucharis and Others

Eucharis grandiflora is a very handsome plant for the warm greenhouse. It has broad shining leaves and numerous erect stems bearing large, intensely fragrant flowers

Amazon and spider lilies are related bulbous-rooted plants with white flowers. The Amazon Lily, *Eucharis grandiflora,* is the most tender and likes a well-warmed greenhouse with temperatures from 15 to 24°C. (60 to 75°F.). The nodding flowers, shaped rather like those of a narcissus, are very fragrant and carried in clusters on 18 to 24-in. stems. It may flower twice or even three times a year, chiefly in winter and spring.

The spider lilies get their name from the long, narrow, spidery segments around the small, funnel-shaped flowers. They belong to two genera, hymenocallis and pancratium and are sometimes listed as ismene. All varieties flower in summer and are nearly hardy.

Pot eucharis bulbs in spring, one bulb in each 6-in. pot, in John Innes No. 2 or peat-based potting compost. Water freely in spring and summer, rather sparingly for a few weeks in autumn, and then more freely again as the plants come into flower.

Spider lilies should be grown in similar soil and 5- or 6-in. pots, but reduce the water supply progressively in autumn and keep the bulbs quite dry from November to February when they are at rest. A minimum temperature of 4°C. (40°F.) is sufficient in winter and little or no artificial heat is required from April to October.

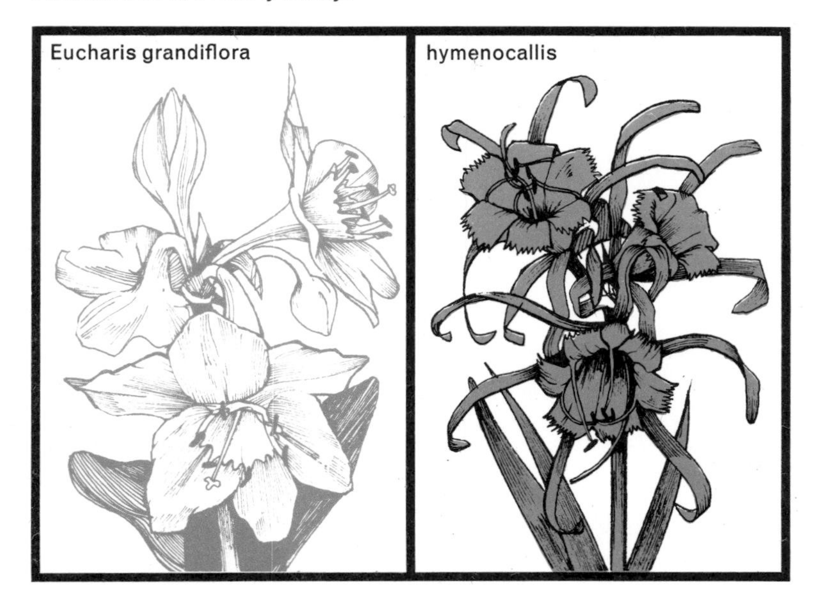

Eucomis and Others

Veltheimias originate in South Africa and make excellent plants for a frost-proof greenhouse. *Veltheimia viridiflora* is the easiest species to grow and it flowers in early spring

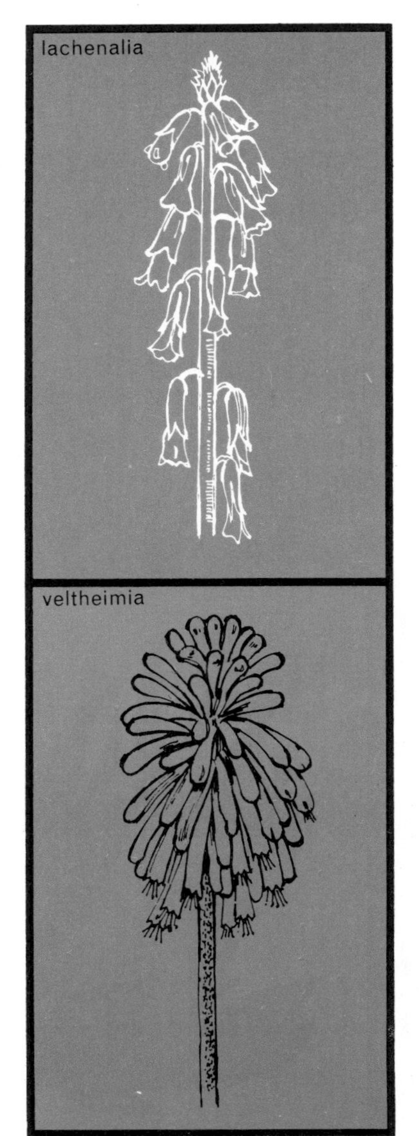

Eucomis punctata is known as the Pineapple Flower. It has strap-shaped, brown-spotted leaves and 18-in. spikes of pale green flowers in July-August. *E. undulata* is similar, but has a tuft of leaves at the top of the flower spike.

Lachenalias are South African bulbs with 1-ft. high spikes of hanging, bell-shaped flowers, mostly orange or yellow though occasionally pale blue, in late winter and spring. They have acquired the popular name of Cape cowslip, though the resemblance to a true cowslip is slight.

Veltheimias are also South African bulbs flowering in winter and spring and the tubular pink or reddish flowers are clustered in little heads like red-hot pokers.

All these plants make excellent pot plants for frost-free greenhouses. Pot the bulbs of lachenalia and veltheimia in August or September, five or six bulbs in each 5-in. pot, in John Innes or peat-based potting compost and grow in a light, airy greenhouse with a minimum temperature of 7° C. (45° F.). Water rather sparingly at first, but more freely once the leaves appear. Gradually reduce the water supply after flowering and keep the plants quite dry in July before re-potting them in August. The pots can stand in a frame or a sunny place out of doors during the summer.

Pot eucomis bulbs in February or March, water freely in summer and keep dry from November to February.

Freesias

It is possible to have freesias in flower throughout the winter and spring by starting batches in succession. These charming plants are quite easy to grow

The little funnel-shaped flowers of freesias are carried in curved sprays on slender stems and are highly fragrant. Freesias can be grown either from seed or from corms, and by using both methods it is possible to have flowers from November until April.

Sow the seed thinly in 5-in. pots in February or March in a temperature of 15 to 18° C. (60 to 65° F.). Do not transplant the seedlings but grow on in an airy greenhouse until June and then stand them in a frame or out of doors in a cool, partly shaded place, bringing them back into the greenhouse in late September. Water the plants freely during this period, but after flowering reduce the water supply steadily and keep them quite dry during June and July when the corms can be shaken out and re-potted.

Corms are potted in August, five or six to a 5-in. pot, in John Innes or peat potting compost and are grown on in the same way as seedlings. Later batches of corms can be potted up until November. Dry the plants off after flowering and continue as before.

Specially prepared freesias can be purchased for growing out of doors. Plant in mid-April in a sheltered, sunny place, water in freely and support them with bushy twigs when the flower spikes appear in summer. As a rule these prepared corms will not flower outside a second year.

Gloriosa, Haemanthus

Gloriosas are lovely climbing plants for the warm greenhouse and they are not difficult to grow. The showy, flame-coloured flowers are abundantly produced throughout the summer

Gloriosa, sometimes called the glory lily or the climbing lily, has scarlet and yellow nodding flowers in summer and climbs to a height of 4 or 5 ft.

Pot the tubers from January to March, one in each 6-in. pot or three in an 8-in. pot, in J.I.P. No. 2 or peat-based potting compost. Water rather sparingly at first, then freely as the plants start to grow. Keep in a temperature of 15 to 21° C. (60 to 70° F.), shading only from strong, direct sunshine, and provide canes for the plants to climb on. After flowering, gradually reduced the water supply and store quite dry in winter in a temperature of 13° C. (55° F.). If little artificial heat is available do not start the tubers before March.

Haemanthus, also known as the blood lily, has wide fleshy leaves and globular heads of starry crimson, scarlet or white flowers in summer. Pot the bulbs in February or March in J.I.P. No. 1 or peat potting compost and grow in a sunny greenhouse without shading, watering moderately at first, freely as growth begins. Keep a minimum temperature of 7° C. (45° F.) but artificial heat is unlikely to be required from April to October as the plants are nearly hardy. Reduce the water supply after flowering and from November until restarting keep the plants dry in a frost-proof place. Do not re-pot every year as these plants flower most freely when rather pot-bound.

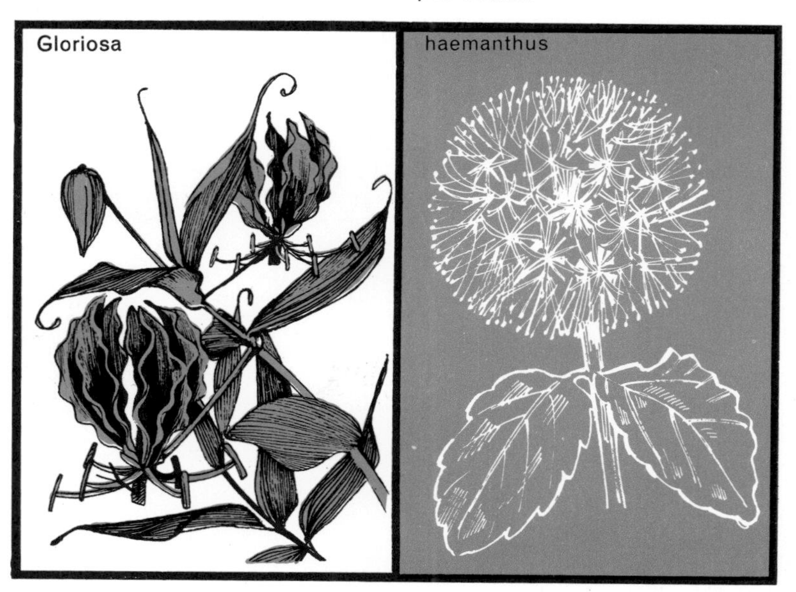

Hippeastrum

The spectacular hippeastrums normally flower in the spring and summer and they are available in white, pink and red. However, specially prepared bulbs can be bought for Christmas flowering

The large, funnel-shaped red, pink or white flowers of hippeastrums are carried on stout 2-ft. stems in spring and early summer and are extremely showy. Hippeastrums are also known as Barbados lilies and amaryllis, but since this last name belongs to a quite different plant, the Belladonna Lily, it can be confusing to use it. They are grown from bulbs which can be purchased in winter or spring.

Pot bulbs singly in John Innes No. 2 Potting Compost in successional batches from January to March to extend the flowering season. Place one bulb in each 5-, 6- or 7-in. pot, only half burying the bulb in the soil. Grow in a light greenhouse or a sunny window in a temperature of 15 to 18° C. (60 to 65° F.), water freely and feed occasionally with weak liquid manure or seaweed extract but do not shade at any time. In September, reduce the water supply progressively and from November until the bulbs are restarted keep them quite dry. Only re-pot them every third or fourth year as hippeastrums flower most freely when their roots are rather crowded. When the bulbs are not re-potted, scrape off an inch of top soil and replace this with fresh potting compost.

Specially prepared bulbs for Christmas flowering can be purchased in the autumn. Pot them as soon as available and grow in a temperature of 21° C. (70° F.).

Lilies

Lilium speciosum rubrum flowers in late summer and autumn. As they are stem rooting, the bulbs should be placed low down in the pots to allow room for a topdressing as the stems develop

Almost any lily can be grown in the greenhouse, but most popular for the purpose are *Lilium longiflorum*, with trumpet-shaped fragrant white flowers in spring and early summer; *L. formosanum*, rather similar in flower but a much taller plant and flowering in late summer; *L. auratum*, with very large, richly fragrant, bowl-shaped white flowers splashed with gold and spotted or banded with crimson in some varieties, also late summer flowering and *L. speciosum*, with nodding reflexed flowers, white flushed with pink or red, produced in late summer or autumn. There are several varieties.

All can be grown in unheated greenhouses but a little artificial warmth in late winter and spring will make *L. longiflorum* flower earlier.

Pot all these lilies in autumn or winter as soon as bulbs can be obtained. Use John Innes No. 1 or a peat-based potting compost without any lime or chalk. Keep them in a cool but frost-proof place and water sparingly until growth starts, then water freely, and grow on in a temperature of about 15° C. (60° F.), shading from direct sunshine. Tie the flower stems to canes pushed well into the soil and after flowering gradually reduce the water supply. When the leaves have died, keep the bulbs quite dry until they are re-potted in the autumn or early winter.

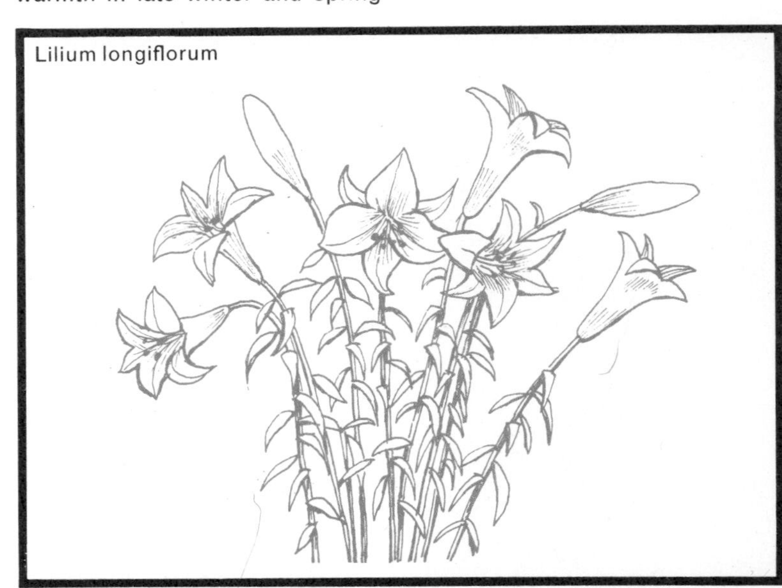

Lily of the Valley, Tuberose

The very fragrant flowers of the Tuberose can be produced at almost any time of the year by potting the bulbs successionally and growing them in a warm greenhouse

polianthes

Lily of the valley or convallaria is not really a bulb but a fleshy crown. Specially selected strong crowns can be purchased in autumn or early winter for forcing into early bloom. Plant these crowns almost side by side in pots or boxes filled with moist peat and keep them in a warm, dark, moist place, such as under the greenhouse staging, until new growth is 3 to 4 in. high, then lift them on to the greenhouse staging. Maintain a temperature of 10 to 15° C. (50 to 60° F.), and keep the peat constantly moist. Cut the flowers as they open and then either discard the plants or harden them off in a frame and plant out of doors in a partially shaded place. Do not force again for at least three years.

The Tuberose or polianthes is a Mexican plant with spikes of intensely fragrant white flowers which can be produced at almost any time of the year by potting the bulbs successionally and growing them in a warm greenhouse at 15° C. (60° F.) or more. Pot one bulb in each 4-in. pot (or four bulbs in a 6-in. pot) in J.I.P. No. 1 or peat potting compost and treat as advised for lily of the valley. Bulbs can be placed in a warm, sunny place and gradually ripened off for flowering again, but it is usually more satisfactory to start afresh with new Tuberose bulbs and lily of the valley crowns each autumn or spring.

Nerine, Vallota

Some nerines can be grown outdoors in well-drained soil and a warm, sunny position. In the greenhouse, all kinds bring welcome colour in the autumn months

Vallota speciosa

Guernsey lily is the popular name of nerine, and Scarborough Lily of vallota. Both are excellent late-flowering pot plants grown from bulbs in a greenhouse. One kind of Guernsey lily, *Nerine bowdenii,* with heads of rose-pink flowers in September and October, is also sufficiently hardy to be grown out of doors in sunny, sheltered places. Other Guernsey lilies are available in various colours from pink and mauve-purple to scarlet. The Scarborough Lily has heads of trumpet-shaped scarlet flowers on stout 18-in. stems in August–September.

Pot bulbs singly in autumn in 4- or 5-in. pots and grow in a sunny greenhouse, minimum temperature 7° C. (45° F.). Water moderately at first, fairly freely while the plants are growing in late winter and spring, but keep Guernsey lilies just dry in a warm sunny place during June and July. The Scarborough Lily should never be quite dry, but should be watered rather sparingly in summer until the flower spikes appear, when it can be watered freely.

Do not re-pot annually as these plants flower most freely when the roots are crowded. Feed the plants in spring with weak liquid fertiliser or seaweed extract.

Greenhouse Climbers

In this section are included in addition to the true climbers some shrubby plants which are most conveniently treated as climbers because their stems are long and flexible. However, they have no natural means of support and must be tied to wires, trellis work or something of the kind.

True climbers may attach themselves by means of tendrils, in which case the supports provided for them must be sufficiently thin for the tendrils to twist around them; or they may actually twine around anything convenient which they touch, which means that they can encompass quite large objects such as posts or pillars. A third group may cling with suckers or aerial roots, by which means they can ascend a wall with no other assistance.

If the climbers are relatively small plants they can be grown in pots into which three or four bamboo canes have been thrust to provide support. Many more are vigorous plants which are happiest when planted directly in a bed of soil and given plenty of space to fill, such as the back wall of a lean-to greenhouse or conservatory or the area beneath the rafters where wires have been strained to hold them up. Some climbers of intermediate vigour can be made into handsome shrub-like specimens by training them over a dome-shaped wire support. This was once very fashionable and might well be used more freely again, especially where such plants can be grown in tubs and used effectively out of doors in summer, perhaps for the decoration of a patio or terrace garden.

A little discretion needs to be exercised in introducing permanent climbers to small greenhouses. All may seem fine for the first year, but once the plants become established they may grow rapidly, monopolise space and cut off a great deal of light. Most of the passion flowers (passiflora) have this tendency, and the fine perennial morning glory, *Ipomoea learii*, and the deliciously scented *Jasminum polyanthum* are two other plants strictly for the large greenhouse or conservatory. There is one good way of growing these vigorous climbers which, though seldom seen, is entirely practical. This is to plant the climber inside the greenhouse but to allow it to grow wholly or mainly outside the house. To do this a hole must be left in the side wall or some other convenient part of the greenhouse or conservatory and all growth arising from the roots guided through the hole and trained up the outside. Treated in this way the stems of quite tender plants may survive many winters and, if they do get killed, more stems are likely to grow from the protected roots to take their place another year. This is a similar practice to the one often adopted for grape vines, though for them with the opposite purpose of allowing the roots to be outside and the stems inside.

Allamanda, Aristolochia

The trumpet-shaped flowers of allamanda are produced from April to September. This climber may be grown in a large container or in the greenhouse border

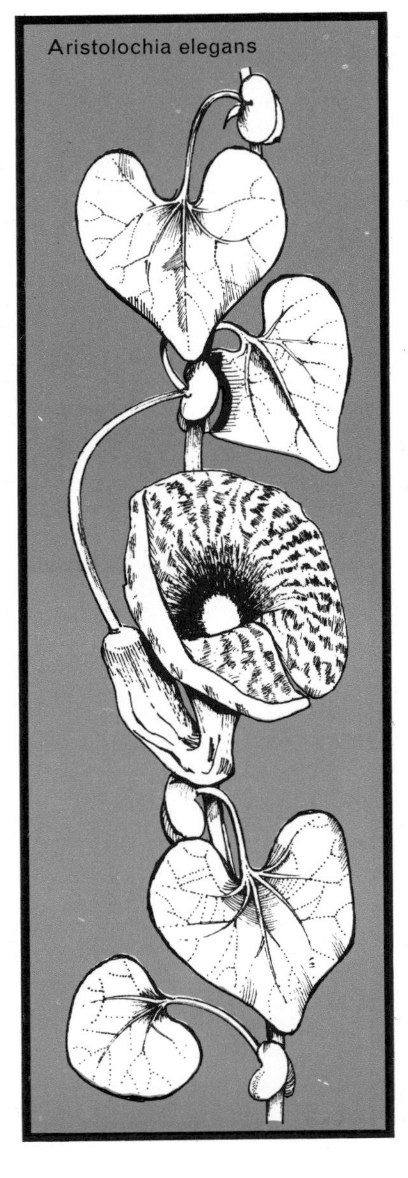

Allamanda cathartica is one of the most handsome climbers, a vigorous plant with shining yellow trumpet-shaped flowers produced from April to September. It can be grown in an intermediate or warm greenhouse in large pots or tubs filled with J.I.P. No. 2 compost or, better still, planted in a border of good loamy soil with some peat and a sprinkling of John Innes base fertiliser. Its long slender stems need to be tied to wires or other supports. It should be watered normally and shaded from strong sunshine. The previous year's growth can be shortened severely each February and in summer the tips of any shoots that extend too far can be pinched out. Propagation is by cuttings of firm young shoots in spring in a well warmed propagator.

Aristolochias are vigorous twiners with curiously shaped flowers like curved funnels. One hardy kind, *Aristolochia durior*, is known as the Duchman's Pipe, an allusion to the shape of its flower. But more attractive for the intermediate greenhouse is *A. elegans*, the Calico Flower, with large, wide-mouthed flowers, white mottled and netted with purplish-brown. It flowers in August and can be grown in large pots or be planted in a bed of soil as for allamanda. Water freely in spring and summer, sparingly in autumn and winter and shade in summer. Growth can be shortened, if overgrown, in February. Increase by early summer cuttings in a propagator.

Bougainvillea, Plumbago Cassia and Others

Bougainvillea makes a splendid cool greenhouse climber and the glorious blossom provides a magnificent display for most of the summer. The colour range includes magenta, pink and orange

Cassia corymbosa is a colourful shrub for a frost-proof greenhouse and the long branches also lend themselves to being trained against a sheltered sunny wall

Both bougainvillea and plumbago are vigorous and showy climbing plants which grow freely out of doors around the Mediterranean, but will not survive frost. They can be grown successfully in greenhouses with a minimum winter temperature of 7°C. (45° F.) and need little or no artificial heat from May to September. Bougainvilleas have magenta, rose, pink or orange flowers and plumbago has pale blue flowers. Both are in flower for most of the summer.

Pot in March or April in 8- to 10-in. pots in J.I.P. No. 2 compost or plant in a bed of good soil to which some peat and bonemeal has been added. Water freely in spring and summer but rather sparingly in autumn and winter. These plants should be grown in full sunlight with plenty of ventilation in summer. Feed the plants once a week from May to August with weak liquid manure and train the stems to wires attached beneath the rafters, or up the back wall of lean-to greenhouses. In March, shorten all sideshoots to a few inches and remove weak or overcrowded stems altogether. Then let the greenhouse temperature rise a little, water more freely and restart the plants into growth. They can be increased by cuttings of firm young growth in summer in a propagator.

Cassia corymbosa is a sprawling shrub, the long flexible stems of which can be trained against a wall, around a pillar or to wires strained beneath the rafters. It is evergreen and produces its clusters of yellow flowers very freely in late summer and autumn.

Most clematis grow best out of doors, but *Clematis indivisa* (*C. paniculata*) is rather tender and grows well in a cool greenhouse. It is evergreen and has white flowers in May and June. *C. florida bicolor* (or *sieboldii*) has white flowers with a central ring of narrow purple segments and is another kind that appreciates shelter, and some hybrids with large double flowers are seen to greatest perfection under glass.

Coronilla glauca is a bushy evergreen which can readily be trained against a wall. It can be grown outdoors in sheltered places but is safer in a greenhouse, and with a little warmth will produce, most of the year, small yellow pea flowers pleasantly scented by day.

All these plants can be grown in any frost-proof greenhouse, but will flower most reliably in a cool house. Grow in large pots or tubs in J.I.P. No. 2 compost or plant in a bed of good loamy soil. Water normally, and lightly shade clematis in summer. Prune early-flowering clematis after flowering, and coronilla in spring as necessary to fill available space. Increase all by summer cuttings in a propagator, cassia also by seed in a temperature of 15 to 18° C. (60 to 65° F.).

Cestrum, Fremontia Cobaea and Others

Fremontias are tender shrubs which must be given frost protection in winter. They flower freely in summer and need well-drained soil and a position in full sun

Cestrums are slender stemmed shrubs best treated as climbers in the greenhouse where they can be trained against walls or up pillars. They produce clusters of hanging, tubular flowers, orange-yellow in *Cestrum aurantiacum*, red in *C. elegans* (also known as *C. purpureum*) and *C. newellii*, yellow in *C. parqui*. All these flower in summer and can be grown in a cool greenhouse either in large pots in J.I.P. No. 2 compost or, preferably, planted in a bed of loam with some peat and sand and a sprinkling of John Innes base fertiliser. Water normally and shade lightly from strong direct sunshine in summer. Prune in February by shortening the previous year's growth. Increase by cuttings of firm young growth in summer in a propagator, temperature 18° C. (65° F.).

Fremontia californica and *F. mexicana* are very similar, evergreen shrubs with long pliable branches which can readily be trained against a wall or on a trellis. The leaves have a spicy odour, the flowers are yellow, saucer shaped and freely produced all summer. Fremontias can be grown in any sunny frost-proof greenhouse or conservatory in just the same way as cestrums, except that they require no shade but benefit from all the sunshine available. If overgrown prune lightly in the spring. Increase by seed sown in spring in a temperature of 15 to 18° C. (60 to 65° F.) and pot the seedlings singly in small pots in J.I.P. No. 1.

Thunbergia alata is better known to many as Black-eyed Susan. It succeeds best when treated as an annual and it makes a splendid plant for a hanging basket

Cobaea scandens is a fast growing perennial but so readily raised from seed that it is often grown as an annual. The purple or white flowers are cup shaped with a saucer-like green calyx behind each and the plant is known as the Cup-and-saucer Vine. It can be grown in any frost-proof greenhouse or even in an unheated house if the plants are purchased in spring and discarded in autumn. Seed should be sown in March in J.I.S. or peat-based seed compost in a temperature of 15 to 18° C. (60 to 65° F.), the seedlings being potted singly in J.I.P. No. 1 and later moved on into larger pots in J.I.P. No. 2 or planted in a bed of good soil. Water freely in spring and summer and, if plants are retained, rather sparingly in autumn and winter. Do not shade. Cut back the vines each February.

Eccremocarpus scaber is also a perennial sometimes treated as an annual. The tubular orange or orange-red flowers are produced from July to October and like the cobaea it climbs vigorously supporting itself by tendrils. Cultivation is similar but plants should be kept almost dry in winter.

Thunbergia alata is a slender twiner known as Black-eyed Susan because each orange, buff or white flower has a black centre. It is usually grown as an annual from seeds sown 3 or 4 in each 3-in. pot in J.I.S. or peat-based seed compost and germinated as for cobaea. Later re-pot the seedlings in 6-in. pots and J.I.P. No. 1.

Dipladenia and Others

Dipladenia is a delightful climber for the heated greenhouse, flowering throughout the summer. It should be shaded from hot sunshine and kept in a fairly moist atmosphere

Dipladenia, mandevilla and trachelospermum are related twiners, the first two climbing readily, the third needing some help. Dipladenias are evergreen and have quite large funnel-shaped flowers, white in *D. boliviensis* and pink in *D. splendens*. There are also garden varieties with deep rose or crimson flowers. *Mandevilla suaveolens*, Chilean Jasmine, is deciduous with white, fragrant flowers. *Trachelospermum jasminoides*, Chinese Jasmine, is also sweetly scented. Its white flowers are small but numerous, its leaves evergreen and leathery. *T. asiaticum* is similar. All are summer flowering.

Pot them in spring in J.I.P. No. 2 or peat-based compost in large pots or plant them in a bed of good soil to which some peat and sand and a sprinkling of John Innes base fertiliser have been added. Grow the dipladenias in a warm or intermediate house, the mandevilla and trachelospermums in any frost-proof greenhouse. Water normally. In summer, shade and maintain a fairly moist atmosphere.

Cut back the previous year's growth of dipladenia and mandevilla nearly to the base in February and in summer pinch out the tips of any shoots that are growing too long. Trachelospermum only requires some thinning in February.

Increase all these plants by cuttings of young growth in summer, rooted in a propagator with a temperature of 18 to 21° C. (65 to 70° F.) for dipladenia, the most tender.

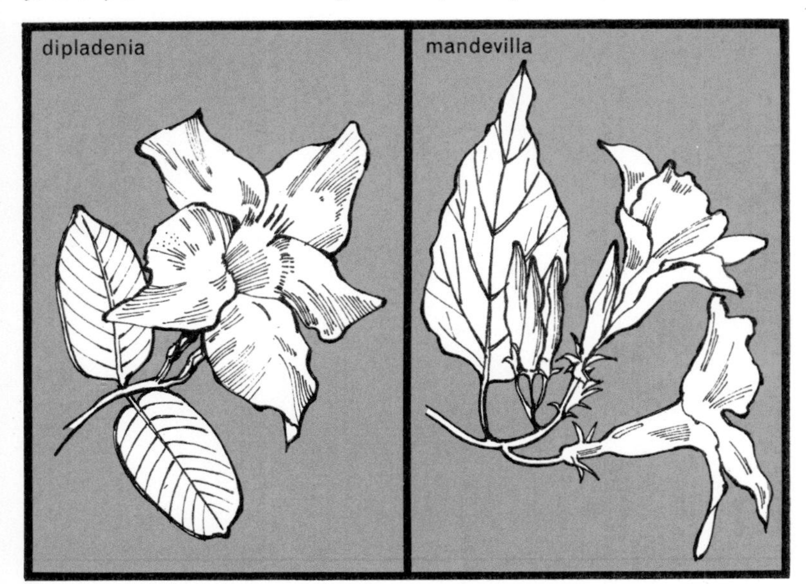

Fuchsia, Pelargonium

This display is formed from varieties of zonal pelargoniums which have been trained up the wall of a lean-to greenhouse. Among those used are Penny, Dryden, King of Denmark and The Speaker

Some varieties of fuchsia with long whippy stems can be trained as climbers either around a pillar or up a back wall of a lean-to greenhouse or conservatory. Vigorous species such as *F. arborescens, F. boliviana, F. cordifolia, F. corymbiflora* and *F. fulgens* can also be grown in the same way. All can be grown in large pots or tubs in J.I.P. No. 2 compost or, better still, planted in a bed of good loam with the addition of peat and sand and a sprinkling of John Innes base fertiliser. They need a cool house temperature and should be watered freely in spring and summer, moderately in autumn, sparingly in winter. They can be fed every 10 to 14 days in summer with weak liquid manure and should be shaded in summer from direct sunshine only. Tips of badly placed shoots can be pinched out at any time and, if necessary, plants can be pruned more severely in February.

Ivy-leaved pelargoniums are naturally trailing plants which can readily be trained up walls, trellis work etc. Some vigorous zonal-leaved varieties can also be trained in a similar way, though they will take longer (perhaps several years) to reach a height of 5 or 6 ft. and will require a good deal more summer pinching and training to get growth where it is required. Soil and culture are as for fuchsias, but no shading is required at any time.

Propagation of both fuchsias and pelargoniums is described elsewhere (pages 78 and 84).

Hoya, Stephanotis

Hoya bella is a trailing plant suitable for a hanging basket and flowering during the summer. It grows better in a higher temperature than is needed for *H. carnosa* and thrives in a warm greenhouse

Both these beautiful greenhouse climbers have waxen-textured flowers so that hoya is popularly known as the Wax Flower and stephanotis as the Clustered Wax Flower. In fact, though related, they are not much alike as the pink and white flowers of hoya are carried in little flat circular clusters, whereas the pure white tubular flowers of stephanotis hang in irregular clusters. Both flower in summer, *Hoya carnosa* rather late.

Pot *Hoya carnosa* in March in J.I.P. No. 1 or peat potting compost or plant in a bed of good loam with peat and a scattering of bonemeal added. Grow in a minimum temperature of 7°C. (45° F.) rising to 15° to 18° C. (60 to 65° F.) with sun heat. Water freely in spring and summer, rather sparingly in autumn and winter, but never allow the plants to become dry. Shade in summer from strong direct sunshine only. The stems should be trained to wires.

Hoya bella is a smaller, trailing plant suitable for hanging baskets. It should be treated in the same way as *Hoya carnosa*.

Stephanotis should be grown in a similar way but in a higher temperature, especially in winter when a minimum of 13° C. (55° F.) is desirable. Feed the plants once a week from May to August with weak liquid manure and spray them occasionally in summer with malathion to keep down mealy bugs and scale insects.

Ipomoea and Relatives

Ipomoeas, with their wide funnel-shaped flowers, are among the most popular of twining plants. They like a position where they will receive as much sun as possible

These are the plants that many gardeners know as convolvulus or morning glory. They are vigorous twiners with widely funnel-shaped flowers produced freely in summer, and all are sun lovers. Several of the most popular are grown as annuals, among them *Ipomoea purpurea* (sometimes called *Convolvulus major*) with purple, crimson or white flowers, and *I. tricolor* (or *rubro-caerulea*) with sky blue, blue and white or reddish-purple flowers. *I. learii* (or *Pharbitis learii*) has deep blue flowers becoming purple and is a perennial.

Related to these, though considerably different in appearance are *Quamoclit lobata* (or *Mina lobata*) with small curiously shaped tubular red and yellow flowers and *Quamoclit pennata* (or *Ipomoea quamoclit*) with scarlet trumpet-shaped flowers.

Though their names seem confused cultivation of all these plants is simple. All can be raised from seeds sown singly in small pots in spring in J.I.S. or peat seed compost in a temperature of 15 to 18° C. (60 to 65° F.). The annual kinds make good pot plants in J.I.P. No. 1 or peat potting composts and can be given canes or trellis to twine around. The perennial kinds are better planted in a bed of good soil and allowed to grow up a wall or on wires beneath the rafters. All will thrive in cool or intermediate houses. Water freely in spring and summer; perennial kinds sparingly in winter. Do not shade.

Jasmines

The charming flowers of the Primrose Jasmine, *Jasminum primulinum*, are to be seen from March to May. It likes a sunny position and the stems should be trained to a support

There are several beautiful jasmines which are either partially or completely tender. *Jasminum primulinum* has large semi-double yellow flowers produced in spring. It is known as the Primrose Jasmine and is a sprawling shrub rather than a climber, though its long, slender stems can be easily trained to a wall or trellis. *J. polyanthum* produces its clusters of white richly scented flowers freely, mainly from February to May under glass. It is a vigorous twiner which will wrap itself around any convenient support. *J. angulare* has white fragrant flowers all summer; *J. sambac* is an evergreen twiner with fragrant white flowers almost throughout the year; *J. rex* has the largest flowers of all, white and up to 2 in. across and *J. grandiflorum* has masses of white sweetly scented flowers in summer.

All these jasmines are best planted in borders of good loamy soil, though they can be grown in large pots or tubs in J.I.P. No. 2 or equivalent compost. *J. primulinum*, *J. polyanthum* and *J. angulare* are happy in any frost-proof greenhouse and are excellent climbers for a cool conservatory. The other kinds are grown in intermediate or warm houses. All require normal watering and should be lightly shaded in summer. The warm-house kinds should also have a moist atmosphere in summer. All can be pruned after flowering as necessary to keep them within bounds. All are readily increased by cuttings or layers.

Lapageria, Bomarea

Lapageria is a beautiful climber for a slightly heated greenhouse or a shaded, sheltered wall and it thrives in a lime-free loam. The waxy bell-like flowers appear in the summer

Lapageria rosea is a slender, evergreen twiner with large, hanging bell-shaped flowers in summer. The flowers may be rose or white and have an almost wax-like texture. The plant is known as the Chilean Bellflower. It is nearly hardy and can be grown in any completely frost-proof greenhouse either in large pots in J.I.P. No. 1 or peat-based potting compost, or planted in a bed of good lime-free loam and peat, with a sprinkling of John Innes base fertiliser. Water freely in spring and summer, moderately in autumn, sparingly in winter, and shade from May to September. Feed every 10 to 14 days from May to August with weak liquid fertiliser. The plant needs a trellis or wires for support. No pruning is normally needed. Increase by layering.

Bomareas are twiners related to the Peruvian Lily (alstroemeria) and with similar flowers. One of the best for cool greenhouse cultivation is *Bomarea caldasii* in which the orange-coloured flowers are produced in large clusters in winter and spring. Cultivation is very much the same as for lapageria, except that there is no need to avoid lime in the soil, summer shading is not essential and the resting period, when little water is required, is in autumn. Plants can be raised from seed sown in a temperature of 18° C. (65° F.) in spring, seedlings being grown singly in small pots. Alternatively, divide the roots in spring.

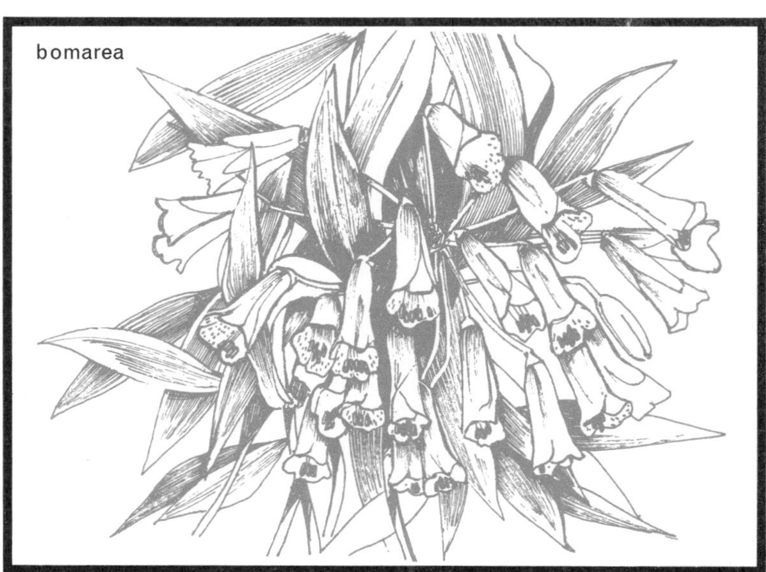

Maurandia and Others

Maurandia erubescens, with its dainty leaves and trumpet flowers, makes a charming climber for a cool greenhouse. It should be grown in a fairly rich compost, such as J.I.P. No. 2

Maurandia are slender climbers supporting themselves by tendrils and producing trumpet-shaped flowers from spring to autumn. Two kinds are commonly grown, *Maurandia barclaiana*, rose and sometimes white and *M. erubescens*, pink and white.

Rhodochiton atrosanguineum (*R. volubile*) is a twiner with small lilac-pink and dark maroon flowers dangling from the stems like tiny peg dolls. It is known as Purple Bells and flowers in summer.

Senecio mikanioides (*S. scandens*) belongs to the same family as the groundsel and ragwort. It is a vigorous twiner with shining, evergreen, ivy-like leaves and clusters of golden-yellow daisy-like flowers in autumn. It is sometimes grown as a house plant for its foliage, but is unlikely to flower unless it gets a lot of light.

All these plants can be grown in a cool greenhouse either in large pots in J.I.P. No. 2 compost or planted in a bed of good loam with some peat and sand and a sprinkling of John Innes base fertiliser. Watering is normal except that the rhodochiton should be watered very sparingly in winter. All can be pruned in spring if necessary.

Rhodochiton is so readily raised from seed sown in spring in a temperature of 15 to 18° C. (60 to 65° F.) that it is sometimes treated as an annual. Maurandias can be increased by seed or by cuttings in a propagator, which is also the best method of increasing the senecio.

Rhodochiton atrosanguineum

Passiflora

All the passion flowers are remarkable and strangely beautiful. *Passiflora allardii* has fine flowers produced in summer, and there are many others to choose from

Passiflora caerulea

The passifloras are known as passion flowers because the circular flowers, often decorated with a ring of filaments and with central, cross-like organs, were thought to represent the Passion of Christ. All are vigorous climbers with tendrils, and flower in summer.

Passiflora caerulea, with blue and white or all white flowers, is one of the hardiest and can be grown out of doors in very sheltered places. *P. allardii* is similar, but has even finer flowers with some pink as well as deep purplish-blue and white. *P. antioquiensis* has large hanging flowers shaped like parasols, rose red in colour. *P. quadrangularis*, which produces the fruit known as grenadilla, also has large flowers, white, heavily veined and netted with purple and fringed by long hanging filaments. *P. edulis* has white and purple flowers then egg-shaped edible fruits.

All can be grown in large pots or tubs in J.I.P. No. 1 compost, but are easier to manage in a bed of good loam. Most are happy in either a cool or intermediate house, but *P. edulis* and *P. quadrangularis* prefer the latter. All should be watered freely in spring and summer, sparingly in autumn and winter, and need never be shaded. They can be cut back to within a foot or so of the base each spring or they can be permitted to retain main stems to which the side growths are shortened. Increase is by summer cuttings in a propagator or the species from seed in a temperature of 18° C. (65° F.).

Solanum and Others

Tibouchina is a very rewarding climber for the cool greenhouse and produces its striking flowers throughout the summer and well into the autumn. They are complemented by the velvet-textured leaves

Solanum jasminoides is a vigorous twiner with evergreen leaves and sprays of slate-blue and yellow (or white and yellow) flowers from July to November. It needs plenty of room though it can be pruned fairly severely each spring. *S. crispum* has purplish-blue and yellow flowers in late summer, but is a sprawling shrub or scrambler rather than a true climber, so its long flexible stems should be tied to suitable supports. *S. wendlandii* has larger flowers and is bushier.

Sollya heterophylla is a slender twiner bearing clusters of nodding sky-blue flowers in late summer and autumn. It is known as the Bluebell Creeper, more from the colour than the flower shape.

Tibouchina semidecandra is a vigorous sprawling shrub, its long lax stems being readily trained against a wall or to wires strained beneath the roof. It has velvety leaves and clusters of fine violet-purple flowers produced all the summer and most of the autumn.

All these climbers can be grown in a cool greenhouse either in pots in J.I.P. No. 2 or peat-based potting compost or in a bed of good loam, with plenty of peat and sand mixed in for the sollya. Do not shade at any time, or only lightly in summer for the tibouchina if its leaves show signs of scorching. Prune in February as necessary to fill available space, and pinch tips of tibouchina as necessary in summer. Increase by cuttings in spring or summer in a propagator.

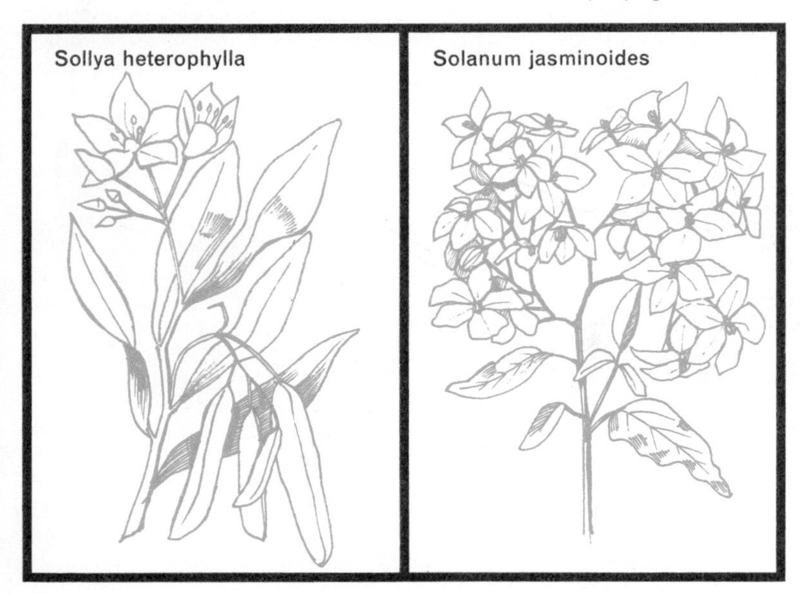

Trumpet Vines

Many related plants are known by the collective name of Trumpet Vine. *Campsis radicans*, shown above, is one of the most useful for providing fast cover

The botanical name of the trumpet vines is rather confused, but here are included climbers variously known as bignonia, campsis, doxantha, pandorea, podranea, pyrostegia and tecoma. All produce handsome clusters of funnel- or trumpet-shaped flowers in late summer and autumn, usually orange-red, orange, salmon red or pink. Some are twiners, a few, including *Campsis radicans*, support themselves by aerial roots like an ivy, and some are scramblers which need a little tying to take their long flexible stems where they are required. All need a good deal of room.

All trumpet vines can be grown in a cool greenhouse either in large pots or tubs of J.I.P. No. 2 compost or in a bed of good loamy soil. All are sun lovers and should not be shaded at any time because unless their growth is well ripened no flowers will be produced. They should be watered freely in spring and summer, moderately in autumn, sparingly in winter, and may be fed fortnightly from June to September with a high potash liquid fertiliser. All can be cut back severely in February and can be increased by cuttings in summer or by layering.

Recommended kinds are *Campsis grandiflora* with large, deep orange and red flowers; *C.* Madame Galen, salmon red; *Bignonia capreolata*, red and yellow and *Podranea ricasoliana*, pink.

Greenhouse Chrysanthemums

The flowering time of chrysanthemums is controlled partly by temperature and partly by night length. It is possible to manipulate both and so produce chrysanthemums throughout the year, and commercial producers of cut flowers and pot plants do exactly that. For the private gardener it is usually more convenient to allow the seasons to take their natural course and to grow the greenhouse chrysanthemums to flower from October to December or January.

There are a great many varieties which are classified according to the shape and size of their flowers and are also divided into Early Flowering, October Flowering and Late Flowering according to their natural flowering season. It is only the last two groups that need to be flowered in greenhouses, but all classes are usually over-wintered in frost-free greenhouses, are increased in cool greenhouses and grown on with frost protection at least until May.

The main flower type divisions are as follows: Single, up to five rows of petals and a button-like central disc; Incurved, petals curling inwards to form a ball-like flower; Intermediate, inner petals curling inwards and outer petals curling outwards; Reflexed, all petals curling outwards; Anemone-centred, like singles but with a low cushion of very short petals in place of a central disc; Thread-petalled or Rayonnante, petals rolled lengthwise like thin quills; Spoon-petalled, petals rolled for part of their length but open at the ends like little spoons; Charm, many small flowers on a bushy plant; Cascade, similar to Charm but with a looser habit.

All chrysanthemums are nearly hardy and do not need a great deal of warmth at any time. They must be protected from frost and also, in winter, from excessive moisture, which can do them just as much injury. From late May or early June until the end of September all chrysanthemums are better out of doors than under glass. In consequence the chrysanthemum routine fits in very well with that for tomatoes which can take their place in the greenhouse in summer, or they may be used with begonias, gloxinias or summer-flowering annuals which do not make much demand upon greenhouse space until about May.

Although greenhouse chrysanthemums are usually grown in pots, most varieties with normal culture grow too tall to make good decorative plants to use indoors. Their proper place is in the greenhouse or conservatory where there is plenty of room for display or to provide cut flowers. The dwarf pot plants that are available in florists' shops and at garden centres are produced partly by treatment with dwarfing chemicals, and partly by growing in controlled day lengths, so that plants can be brought into bloom before they have made much growth. However, there are a few naturally dwarf varieties such as Blanche du Poitou, Blanche Poitevene, Marie Morin, Yellow Morin, Dorothy Wilson and Dwarf Rose. These should be grown from April or May cuttings to flower from September to November.

Propagation

Anemone-flowered chrysanthemums, with their distinctive cushion of florets in the centre of the bloom, are ideal for flower arrangements. They are available in many different shades

All chrysanthemums can be raised from seed but usually this method is only used for the Charm and Cascade varieties as others give variable results. The seed is sown in February or March in a temperature of 15 to 18° C. (60 to 65° F.) in J.I.S. or peat-based seed compost. Seedlings are pricked out $1\frac{1}{2}$ in. apart in J.I.P. No. 1 or peat-based potting compost and are then treated in the same way as plants raised from cuttings.

Cuttings are prepared between January and May from young shoots, 2 to 3 in. long, growing from the roots of old plants which have been cut hard back after flowering and kept moderately moist in a temperature of 7 to 10° C. (45 to 50° F.). Each cutting is cut off cleanly just below a joint, the lower leaves, if any, are removed and the base is dipped in hormone rooting powder. The cuttings are inserted in a mixture of equal parts coarse sand, loam and peat in pots or shallow boxes. They are rooted either in a propagator or on the open staging of the greenhouse, in which case they will need rather more watering and overhead spraying to prevent excessive flagging. An air temperature of 10° C. (50° F.) should be maintained and the soil should be warmed from below to 15° C. (60° F.).

When rooted, cuttings should be potted in 3- or $3\frac{1}{2}$-in. pots in J.I.P. No. 1 or peat-based potting compost and grown on in a cool greenhouse. Shade for a few days then allow all the light available.

Growing On

The Cascade chrysanthemums are delightful pot plants which can be raised from seed. It should be noted, however, that the plants take up a considerable amount of space

As soon as the small pots are full of roots plants should be moved to 5-in. pots and J.I.P. No. 2 or peat-based compost. A few weeks later a further transfer will be necessary to 7- or 8-in. pots and J.I.P. No. 3 or peat-based compost. At this stage each plant should be provided with one strong 4-ft. cane for support.

Meanwhile the growing tip should be removed from each plant when it is 8 or 9 in. high and the tips of side growths similarly removed when they are 8 or 9 in. long. This encourages good branching from low down but it is not always wise to retain all the branches that result. The number retained will depend on the kind of flowers required. For very large exhibition blooms three stems to each plant may be ample. At the other extreme for a big display of small flowers a dozen or more stems may be required. However many stems are retained each should be looped individually to the central cane to prevent breakage. Very large plants may require several canes angled to lean outwards at the top to allow for the spread of the stems.

In late May (or as soon as danger of severe frost is over) plants should be placed out of doors in a sunny position. Stand the pots on a gravel or cinder base and tie the support canes to wires strained horizontally between posts driven in at the ends of each row of pots. Leave some room between rows.

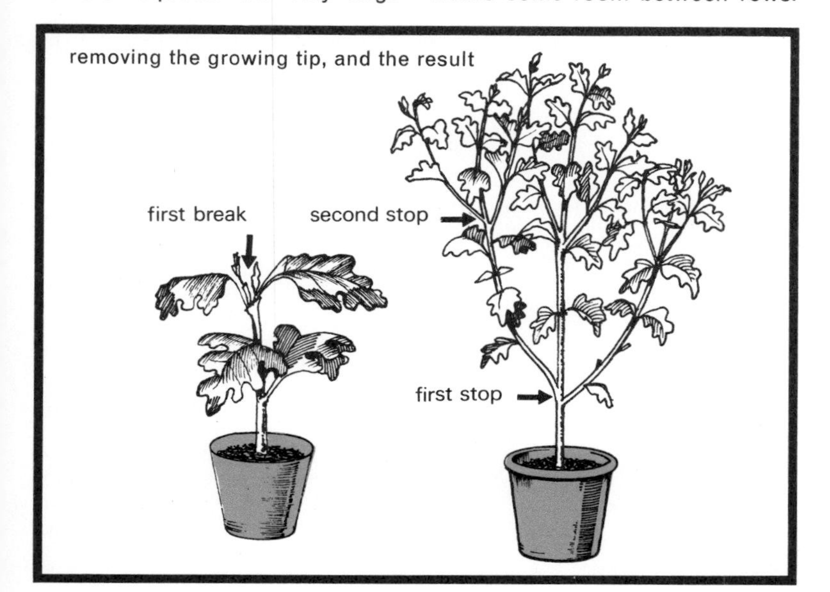

Flowering

The late-flowering chrysanthemum Mavis Shoesmith. This is a good large-flowered incurved variety and is typical of the very popular exhibition type

Throughout their growing season chrysanthemums must be watered freely. From mid-June until the first petal colour is seen in the opening flower buds plants should be fed every 7 to 10 days with weak liquid fertiliser, for preference a special chrysanthemum fertiliser.

When flower buds start to appear it is necessary to decide whether large flowers (one to each stem) are required or sprays of smaller flowers. If the former, only the terminal bud on each stem should be retained, all others being removed at an early stage. If sprays are required the terminal bud is removed and all side buds kept.

Cascades are stood in summer on a raised plank running east/west, each plant with a cane sloping to the north to which the main stem is tied. This cane is gradually lowered, other canes being lashed to it for support, and side growths are pinched.

At the end of September all plants are returned to the greenhouse and grown with maximum light and as much ventilation as is consistent with maintenance of a temperature of 10 to 15° C. (50 to 60° F.). A little artificial heat may be necessary at times to dry the air and prevent decay of the opening flowers. After flowering, stems are cut down and pots kept with frost protection only until it is time to start growth again to provide a new lot of cuttings. Charm and Cascade chrysanthemums are usually discarded after flowering.

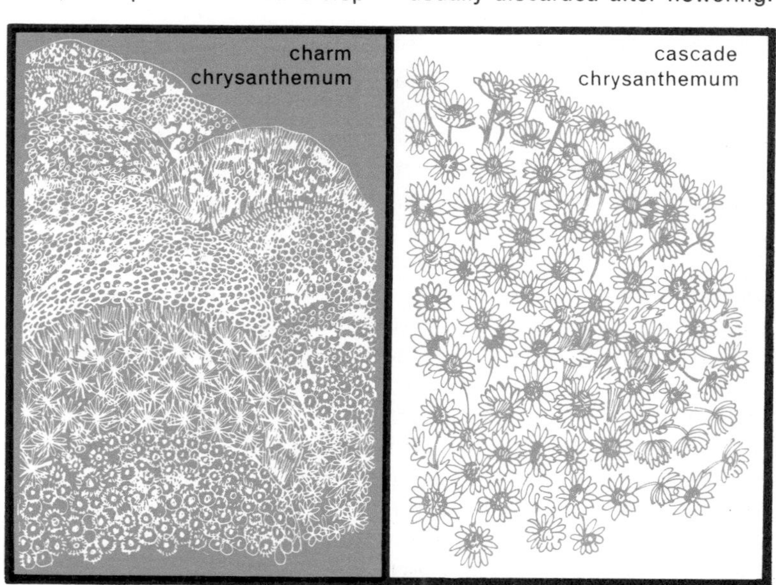

Orchids

The orchid family is a very large one including thousands of distinct kinds growing naturally in many different parts of the world. Some are tropical plants and others, a number of which are native British wild plants, are completely hardy. In addition there are numerous hybrids and varieties.

Many orchids have thickened stems, known as pseudobulbs, which enable them to store food and moisture and rest for quite long periods. Some are epiphytic plants, getting most of their food from the air, but terrestrial orchids root and feed in the soil in the ordinary way. In nature many orchids grow in forests, the epiphytic kinds clinging to trees or damp rocks.

Because of these peculiar methods of growth orchids need composts very different from those used for most other greenhouse plants, and their management is also different. This does not mean that they are difficult to grow, in fact most orchid enthusiasts insist that they are among the easiest greenhouse plants to manage and they will withstand a great deal of ill-treatment. It does mean that a gardener who has been accustomed to handling other greenhouse plants must learn new ways when embarking on orchid cultivation. In particular, since many plants will be growing in spongy mixtures of sphagnum moss and osmunda fibre (perhaps even in completely synthetic materials such as expanded polystyrene if plants or composts are purchased from a commercial grower), normal methods of assessing soil moisture must be abandoned. In place of this the eye must be educated to observe by the look of leaves and composts when water is required. Fortunately, since the compost is so open and spongy it is almost impossible for it to become waterlogged and, since the atmosphere for orchids is usually humid, they do not dry out as rapidly as some other pot plants. Perhaps it is in identifying the resting season, which is a characteristic of many orchids with pseudobulbs, and in keeping them sufficiently dry during this important period in their annual cycle that the greatest difficulty arises. This is largely a matter of knowing the growth pattern of each species or hybrid and since there are many thousands of these it is impossible to deal with them in detail in a book of this character. There are numerous books devoted exclusively to the cultivation of orchids. There are also active orchid societies which publish useful literature on the subject and enable their members to exchange information. Commercial orchid firms are also always ready to assist customers and, because as a rule they grow nothing but orchids, they have an expertise which is unsurpassed.

Though it is possible to grow orchids in a mixed collection of greenhouse plants it is easier to provide the special conditions they require in greenhouses devoted exclusively to them. Houses specially suitable for orchids are produced by some greenhouse manufacturers.

Propagation

The belief that orchids are difficult to grow and expensive should not be a deterrent to the would-be grower, for there are many, such as these cymbidiums, which are fairly easy and inexpensive

Orchid seed is dust-like and produced in great quantity. In nature much seed never germinates and vast numbers of seedlings die young. In nurseries it is possible to germinate seeds in test tubes or flasks on agar (a special kind of jelly), impregnated with nutrient solution with all harmful germs excluded, so that a high percentage of the seed grows and the seedlings survive. This has made possible the immense increase in hybrid orchids, and unflowered seedlings with good parentage can often be purchased quite cheaply. But hybrids usually differ from their parents, sometimes for the better but often not, so purchasing unflowered seedlings is to some extent a gamble.

Divisions and cuttings reproduce exactly the plants from which they were taken. A system of taking minute cuttings from the growing tips of plants and growing them in slowly revolving flasks of nutrient solution has made it possible to increase selected plants far more rapidly. Small plants grown from these meristem cuttings are usually reasonably priced and can be grown on to flowering size in a few years.

Orchids are also still collected in the wild and imported and some firms specialise in selling such plants which can often prove very interesting.

Care of Plants

The three main requirements of most greenhouse orchids, such as this miltonia, are an equable temperature, a moist atmosphere and no direct sunshine. It is useful to remember that orchids prefer rain water.

Most greenhouse orchids need an equable temperature, a moist atmosphere and no direct sunshine. They can be grown in any adequately heated greenhouse, but one with solid walls rather than with glass to the ground is ideal, with slatted wooden blinds fitted outside to run on rollers 8 or 9 in. above the roof glass. Pots should stand on slatted staging about 6 in. above solid staging covered with small pebbles or grit that can be kept constantly moist.

Water the plants, if possible, with rain water. Established plants should be watered fairly freely while they are in growth, moderately for a few weeks after re-potting, and rather sparingly while at rest. Water paths and under staging at least once daily in spring and summer to maintain humidity, less frequently in winter, and syringe plants with water daily in warm weather. Alternatively, a mechanical humidifier can be fitted. Remove resting plants to the coolest part of the house. Shade from early April until the end of September, but if moveable blinds are fitted only use them in spring and early autumn during the middle period of the day when direct sunshine might harm the plants. If solid walls come above the staging level they will provide all the early and late shade necessary.

Potting and Dividing

The genus *Dendrobium* is the largest one in the orchid family. This beautiful species, *D. thyrsiflorum*, is evergreen and the flowers appear in spring

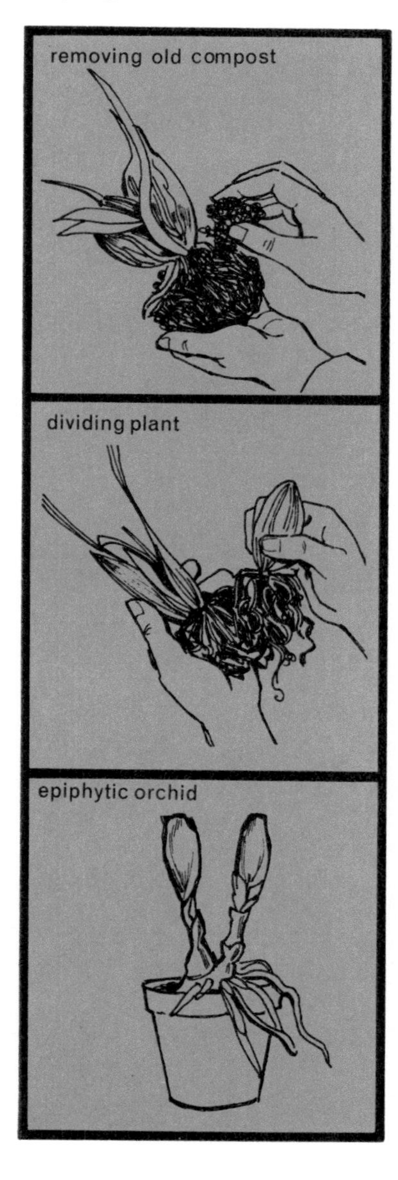

Most orchids can be grown in a mixture of 3 parts by volume osmunda fibre and 1 part chopped sphagnum moss. These materials can be obtained from orchid specialists who often prepare and sell the required mixtures. For some strong-growing or terrestrial orchids the addition of 2 parts fibrous loam may give even better results, but the loam should be passed through a $\frac{1}{2}$-in. mesh sieve and only the soil left in the sieve used, the fine material being discarded.

Use a clay pot just large enough to contain the plant. Place enough pieces of broken pot (crocks) in the pot to fill it to one third, work some compost between the roots of the plant, place it in the pot and fill in the hollows with more compost.

When re-potting old plants remove most of the old compost and then divide the plant, keeping four or five pseudobulbs and a new growth to be potted on, and discard the rest.

As they grow, the epiphytic kinds will push roots out over the side of the pot or develop them in mid-air from stems or pseudobulbs. This is their natural habit and should not be interfered with.

Most orchids only need re-potting every second or third year but slipper orchids (paphiopedilum or cypripedium) may be re-potted every year. Spring is the best time, but not while plants are in flower.

Aerides and Others

Calanthe vestita is a terrestrial orchid from Burma and Malaya. The shining white flowers are carried on spikes up to 2 ft. long, which are produced during the winter

The flowers of aerides are not big but they are numerous, carried in long arching spikes, and they last well. Colours are mainly pink, purple and white. There are many species from different parts of Asia and though all are epiphytes to be grown in sphagnum moss and osmunda or equivalent compost they differ in their temperature requirements. Most can be grown in an intermediate house, but a few need a warm house. All need shade in summer. Water well from May to October then sparingly.

Angraecum sesquipedale is a very distinctive orchid with fan-like clusters of leaves and hanging flower stems carrying large wax-white flowers each with a long slender spur like a tail. All angraecums can be grown in an intermediate greenhouse in any of the composts recommended for epiphytic orchids. Treatment is similar to that for aerides, but with slightly more water in winter as these orchids have no marked resting period.

In contrast *Calanthe vestita* goes completely to rest from about September to February, when it loses all its leaves and can be kept absolutely dry. It is then re-potted in a compost of loam with a little peat, chopped sphagnum and sand. At first it is watered sparingly, then as growth starts much more freely and it is grown in an intermediate house with light shade. The spikes of rose-pink and white flowers appear in winter during the resting period.

Cattleya, Miltonia

Cattleyas make especially good cut flowers and are much used by florists in corsages. Most flower in autumn and winter and the large blooms are brilliantly coloured

Cattleyas and miltonias are among the most gorgeously coloured of the orchids. Cattleyas have very large flowers, often in shades of magenta or mauve with or without white, and make fine cut flowers much in demand for corsages. There are a great many varieties including complex hybrids with other orchids, some of which have produced the most highly coloured flowers. They will grow well in intermediate house conditions and require only light shading. Most flower in autumn or winter and rest for a few weeks afterwards when water should be applied sparingly. For the rest of the year water freely but allow the compost to become almost dry before watering again.

Miltonias have flowers rather like huge pansies, often white, strongly blotched with rose or crimson. They flower in spring and early summer and good plants can be produced in 4- or 5-in. pots. They will grow well in cool house conditions, though the temperature should not drop below 13° C. (55° F.) at any time. They enjoy a fair amount of shade in spring and summer and a humid atmosphere, so should be watered throughout the year, though less frequently in winter than in summer. Annual re-potting is desirable.

Miltonia hybridises with odontoglossum to give odontonias. Like laeliocattleya, brassocattleya and brassolaeliocattleya (from cattleya with laelia and brassavola) they extend the colour range and form.

Cymbidium, Coelogyne

Coelogyne cristata is a very attractive orchid flowering in March and April on pendulous spikes. It needs a cool winter temperature and the roots should not be disturbed

Cymbidium and coelogyne are among the most popular of greenhouse orchids. The flowers, of many different colours, are carried on arching stems and last extremely well when cut. Plants are fairly large and have a tendency to flower every second year.

Cymbidiums are classed as cool house orchids and will survive even in greenhouses in which the temperature falls to 7° C. (45° F.) on occasions. They are epiphytic orchids but do well in composts containing some loam as recommended for terrestrial orchids (page 107). They grow throughout most of the year with a less marked resting season than many orchids so they should never be allowed to become dry.

Coelogyne cristata is a small, spreading orchid with white flowers hanging in short spikes in spring. It will thrive under exactly the same conditions as the cymbidiums with plenty of water in spring and summer. Only re-pot when absolutely necessary as large, old plants flower most freely. Plants can be kept going by replacing some of the old compost from time to time without actually removing them from their pots.

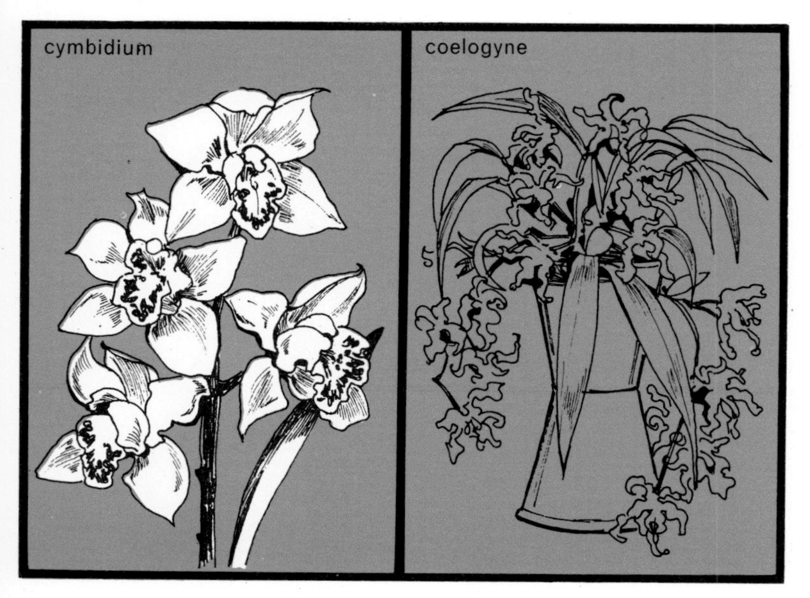

Dendrobium, Vanda

Vanda caerulea is an orchid for the intermediate greenhouse. It should be re-potted very infrequently, but each spring as growth re-starts some of the compost should be replaced

Most useful for the amateur's greenhouse are the hybrids of *Dendrobium nobile* which will thrive under intermediate house conditions. They have long, thin, cane-like pseudobulbs and clusters of white, yellow pink or purple flowers in winter and spring, often strikingly blotched with one colour on another. Hybrids of *D. phalaenopsis,* mainly in shades of rose and magenta, need warm house conditions. Water both groups freely in spring and summer but decreasingly during the autumn until the flower buds appear, when more water should be given. Maintain a humid atmosphere in summer and shade lightly. Every second or third year re-pot after flowering is completed.

Vanda caerulea is very distinctive, with wide sprays of pale blue flowers produced regularly each autumn. It requires intermediate house conditions. Water moderately at all times allowing the compost to become nearly dry before wetting it thoroughly again, and shade lightly in spring and summer. Direct sunshine will scorch plants but heavy shade pales the flowers. Re-pot very infrequently but replace some compost annually as growth re-starts in the spring.

Odontoglossum and Others

The delicate flowers of *Odontoglossum crispum* are carried in slender sprays and are excellent as cut flowers. This orchid also makes a very attractive pot plant

There are a great many odontoglossums covering a wide range of colours and forms, but two types are of particular interest to gardeners. *Odontoglossum crispum* carries its white flowers in slender sprays and is first class both as a pot plant and as a cut flower. There are also a great many hybrids of the crispum type, but with flowers variously splashed with rose purple or maroon. All these need intermediate house temperatures. *Odontoglossum grande,* by contrast, will grow in a cool house and carries its large yellow and cinnamon-brown flowers in small spikes, usually of three. It is an easy orchid to grow, provided it is given a good rest with little water in winter. *O. crispum* and the hybrids have a less marked resting season, need very careful watering at all times, and are not the easiest of orchids to grow well.

Odontiodas are hybrids of *O. crispum* with cochlioda, which introduces red and rose shades.

Oncidiums usually have small yellow or yellow and brown flowers produced in large, branched sprays, but some have larger flowers produced successionally, one or two at a time, for most of the year. One of these, *Oncidium papilio,* is known as the Butterfly Orchid because of the shape of its flowers. This and *O. rogersii* require warm house conditions, but the small *O. varicosum* can be grown in a cool house under similar conditions to cymbidiums. It flowers in autumn and winter.

Slipper and Moth Orchids

The slipper orchids, paphiopedilums, are terrestrial plants with highly distinctive pouched flowers. *Paphiopedilum insigne* is easy to grow in a cool greenhouse or even in a well-lighted room

Paphiopedilums are the slipper orchids still familiar to many people by the old name cypripedium. They are terrestrial orchids with highly distinctive pouched flowers usually carried singly on 1-ft. high stems. They are compact plants easily handled in small greenhouses. *Paphiopedilum insigne* and some kinds closely allied to it can be grown under cool greenhouse conditions, or even in well-lighted rooms, but the larger-flowered hybrids and those with mottled leaves require the greater warmth of the intermediate house. All like a compost containing some loam. They enjoy shade in spring and summer, ample humidity and, as they have no pseudobulbs and no marked resting period, must be watered all the year, but less frequently in winter than in summer.

Phalaenopsis (moth orchids) have white or pink flowers of highly distinctive shape carried in long spikes and produced at various times of the year. They are usually regarded as warm house orchids, but provided they are not overwatered in winter they can be grown under intermediate conditions. They like plenty of moisture in the air and a good deal of shade in spring and summer.

Small Orchids

The genus *Lycaste* has large, long-lasting, waxy flowers–characteristics which make it very popular. Flowering throughout the spring, it will grow in a cool greenhouse

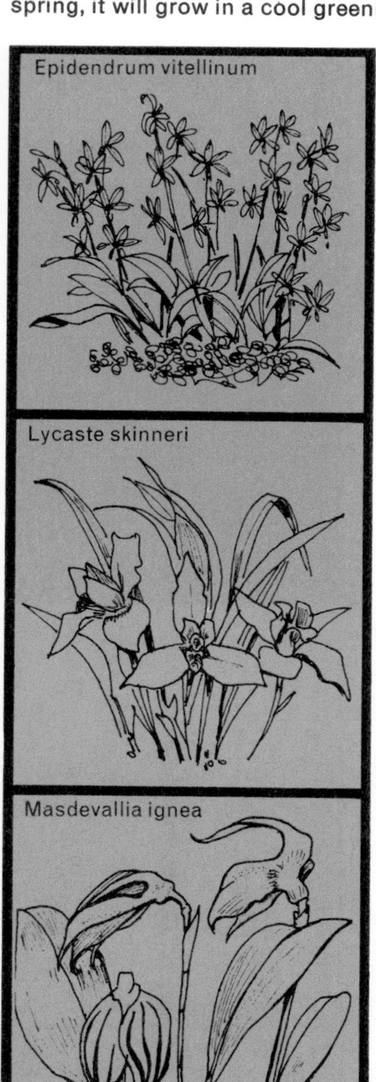

Dendrochilums are very distinctive plants usually with small white or greenish-yellow flowers hanging in long slender chains. They are epiphytes and require a warm house.

Epidendrum vitellinum is a delightful orchid with vermilion flowers produced successionally most of the summer. It can be grown in intermediate house conditions.

Lycaste skinneri, with white or pink flowers on short stems throughout the spring, will grow in cool house conditions with a little loam in the compost. It rests in winter, when it wants little water.

Masdevallia coccinea and *M. ignea* are small but brilliant orchids with purple, crimson or vermilion flowers in spring and early summer. They need cool house treatment and should be watered throughout the year.

Maxillaria is so called because it was thought that the flowers of some kinds resembled the jaws (or maxilla) of an insect. Those of *Maxillaria picta* are yellow and maroon, sweetly scented and produced in winter. It likes cool house conditions and should be watered sparingly in winter.

Sophronitis grandiflora and *Stanhopea tigrina* are both best grown in baskets suspended from the rafters of an intermediate temperate house. Water both throughout the year but sparingly in winter.

Zygopetalum mackayi, with curiously shaped flowers on erect stems in winter, likes an intermediate house and loamy compost.

Pests and Diseases

Florence Stirling, one of the many lovely hybrids of odontoglossum. As with all orchids, a careful watch must be kept for pests and diseases against which there are several effective controls

Orchids are attacked by most of the pests common to greenhouses. Red spider mites cause grey mottling of the leaves and thrips produce brown streaks on the flowers. Scale insects, like minute limpets, adhere to the leaves and exude a sticky fluid on which grows a black mould. Greenflies may cluster on the young shoots checking or distorting them, mealy bugs with their protective white waxy covering secrete themselves on the plants and slugs devour leaves or stems.

Many of these pests can be controlled by sponging the leaves occasionally with water containing a little white oil emulsion. Plants can be sprayed occasionally with malathion or derris or the house may be fumigated from time to time with smoke generators containing BHC. Slugs and snails can be destroyed by placing metaldehyde bait on the staging between the pots.

Diseases are less numerous and, on the whole, less controllable except by good cultivation and hygienic methods, including the removal of all dead or rotting leaves and stems. Occasionally viruses infect the plants causing ring-spotting, brown streaking or distortion. Infection is spread by greenflies and other insects, so if these are kept away trouble will be reduced to a minimum.

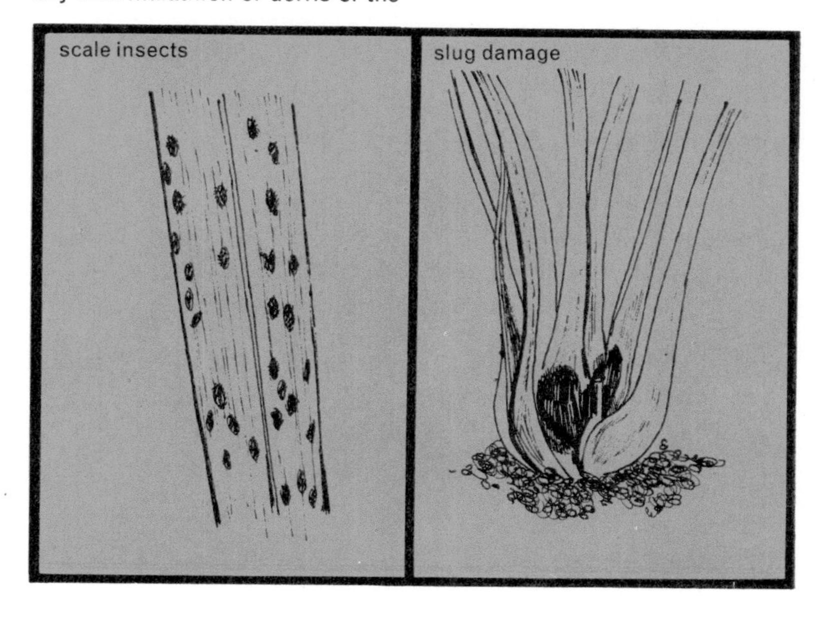

Greenhouse Ferns

Though a few ferns grow naturally in sunny places, most are shade lovers. They are, therefore, an excellent choice for a greenhouse that gets little or no direct sunshine. Some kinds make good indoor plants but many find the air of living rooms too dry for their liking, though if they are given periods of recuperation in a suitably damp place they may be kept in good condition indefinitely. One old-fashioned but still delightful way of reconciling the undoubted charms of ferns for indoor decoration with their love of moisture is to grow them in a conservatory or garden room with direct access from a living room. Here the ferns can be visible at all times and the fernery can be a pleasant place to sit in on hot days.

Another possibility is to grow ferns in rooms but inside special growth cabinets in which the moist conditions they like can be maintained without difficulty; or some small kinds can be grown in Wardian cases or bottle gardens.

Ferns are not flowering plants and their method of reproduction is markedly different from that of flowering plants. They produce dust-like spores in structures known as sori, which appear as brownish or blackish spots or patches on the backs of special leaves or fertile fronds. When these spores are sown under suitable conditions they develop into flat green plants, or prothalli, of a quite different character to the original fern. The prothalli carry male and female organs for fertilisation and production of a new generation of fern plants. Because of this method of sexual increase it is impossible to make the controlled crosses between ferns that are possible with many flowering plants. Instead, if an attempt at hydridisation is to be made between two different fern species, or even between varieties of the same species, spores of the two kinds must be sown together in the hope that when the prothalli develop, cross fertilisation (which is effected by mobile, water-borne male sex cells) will take place and some hybrids will eventually appear.

Some ferns produce little plantlets, like bulbils, on their fronds and if these fronds are pegged to the surface of moist peaty soil without being detached from their plants, the plantlets will develop and can be detached and potted singly. They will resemble their parents in every particular and provide a useful method of increasing some garden varieties with beautiful frond variegations. Division of the rhizomes or creeping stems produces similar results and is best carried out when re-potting in spring.

These frond variegations are not, as a rule, transmitted by spores though they do occur spontaneously, though rarely, by natural mutation or sporting. There are not so many of the frond variegations among the tender as among hardy ferns, probably because there has not been the same effort to search for them and preserve them.

Raising Ferns

Asplenium is an important genus of ferns which is known by the common name of spleenwort. This species, *Asplenium nidus*, prefers a warm greenhouse, though it will survive at lower temperatures

Ferns can be increased most readily by dividing large plants when replanting or re-potting. Take care that each piece contains a growing shoot or crown and roots. Where possible break up the large plants with the hands only, but if necessary use two hand forks thrust in back to back in the middle of the plant and lever them apart.

With plants such as *Asplenium bulbiferum* which make tiny plantlets on the fronds, peg these fronds to the surface of moist soil without removing them from the plant. When the plantlets begin to grow and form roots of their own, cut them out carefully with the soil attached and plant them in 3-in. pots.

To raise ferns from spores use moist peat seed compost in clean plastic pots. Make the surface quite level and moderately firm, dust the spores over it as evenly as possible and cover with a pane of glass. Keep the pot in a shady frame or greenhouse and when water is required give it by holding the pot almost to its rim in water for about a minute, then allow the surplus to drain away.

The prothalli will appear as green scale-like growths on the surface of the compost. Some weeks later tiny fronds will appear and when the little plants are large enough to be handled they can be carefully pricked out into pans or boxes of the same compost as used for potting ferns.

Cultivation

Platycerium bifurcatum is aptly named the Stag's Horn Fern. It can be grown on a block of wood or cork and this can then be suspended so that the fern is growing on its side

Greenhouse ferns should be grown in a mixture of 2 parts by volume sphagnum peat or well-rotted beech or oak leafmould, 1 part medium loam and 1 part coarse sand or well-broken charcoal.

Re-pot when necessary in March but as a rule this need only be done every second or third year. Water the plants rather sparingly for the first few weeks after re-potting then more freely as the roots grow out into the new soil. Water established ferns fairly freely from April to September, and just sufficiently to keep the soil moist for the rest of the year.

Grow ferns in a north-facing greenhouse or in one that is shaded from direct sunshine, if necessary fixing slat blinds on the house so that the strong light is broken.

Ideal temperatures differ according to the kind of fern being grown but most of those commonly cultivated in pots will thrive in a winter minimum temperature of 7°C. (45°F.). It is unlikely that artificial heat will be needed from April to October inclusive when the problem is often to keep the greenhouse sufficiently cool. This will be helped by syringing with water the walls, paths and staging between the pots, but do not wet the fronds of the plants themselves.

Kinds of Ferns

Maidenhair ferns are amongst the most handsome of ferns. *Adiantum venustum* is a splendid greenhouse plant and, since it is almost hardy, it can be planted in sheltered places out of doors

Adiantum, the maidenhair fern, is one of the most popular and beautiful of all. There are numerous kinds, the most frequently grown being *A. cuneatum*. It enjoys warmth and in poorly heated houses may lose its leaves in winter, but new ones will appear in spring. *A. capillus-veneris* is much hardier and can be grown out of doors in some places, while *A. pedatum* is the hardiest of all and is often planted out of doors. It will lose its leaves in winter, but under glass it remains evergreen.

Of the aspleniums, known as spleenwort, by far the most popular is *A. bulbiferum*, so called because of the little plantlets which form like bulbils along the fronds and which readily grow into new plants. Great quantities of this fern are produced commercially as it is one of the best for room cultivation as well as for greenhouses. *Asplenium nidus* is very different in appearance and requirements, making a shuttlecock of long, undivided shining green fronds, and enjoying a fairly warm greenhouse though it will survive at lower temperatures.

Blechnum gibbum makes a plume of fronds on top of a short, thick stem, almost like a miniature tree fern. It is distinctive and easy and can be grown in rooms as well as greenhouses.

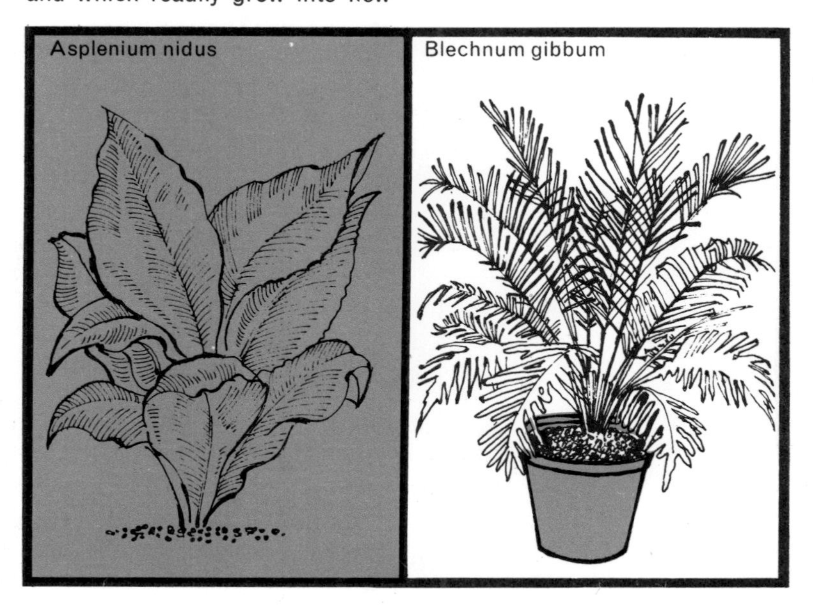

Kinds of Ferns

Woodwardia radicans is an ideal fern for the cool greenhouse or conservatory. It is important to keep it under cool conditions as in heat it is prone to attack from certain pests

Pteris arguta is one of the ribbon ferns and it should be grown in a warm greenhouse. It is an especially attractive fern with delicate, well-marked foliage

Cyrtomium falcatum

Davallia bullata

Cyrtomium falcatum is a popular commercial fern equally suitable for rooms or greenhouses. Known as the Holly Fern, because of the thick texture and dark green colour of the large segments of its fronds, it is very nearly hardy.

Davallia bullata is known as the Squirrel's-foot Fern, and *D. canariensis* as the Hare's-foot Fern, because of the shape of the brown, furry ground-hugging stems from which the finely divided fronds grow. They are among the loveliest of ferns, and there are numerous other beautiful davallias, many of which make good plants for hanging baskets.

Dicksonia antarctica is the Australian Tree Fern, which makes a great cartwheel of fronds on top of a stout, tall trunk. Young specimens can be grown in pots but as they get bigger they need to be planted in borders of soil. In sheltered seaside gardens in the south and west they can be grown out of doors in damp shady places.

Lygodium japonicum is one of the few climbing ferns and should be given bamboo canes or vertically strained wires around which it will wind itself. Its finely divided fronds are very attractive.

Woodwardia radicans can be grown out of doors in a sheltered position provided it is well protected in winter, and it is also an attractive plant for a hanging basket.

Nephrolepis exaltata is known as the Ladder Fern because of the ladder-like pattern of its fronds, which may be 3 ft. or more long. There are a great many varieties in some of which the fronds are so finely divided that they look like moss. All are easily grown in rooms or greenhouses.

Platycerium bifurcatum, known as the Stag's Horn Fern because of the antler-like appearance of its grey-green fronds, does not require soil and will grow well on a block of wood or cork to which it can be bound with a wad of sphagnum moss around its roots. The block can then be suspended so that the fern is growing on its side, which is its natural habit. Water the plants by immersing them for a few minutes, daily in spring and summer, much less frequently in autumn and winter. This fern can be grown in rooms.

Some of the greenhouse kinds of pteris are known as ribbon ferns because of the long, narrow, ribbon-like segments to the fronds. They are among the most popular of pot ferns because of the ease with which they can be grown in greenhouses or rooms. The principal kinds are *Pteris cretica* which has a variegated variety with a white stripe down the centre of each frond segment, and *P. serrulata* which has a lovely variety with crested fronds. *P. tremula* has more finely divided fronds and is equally easy to grow.

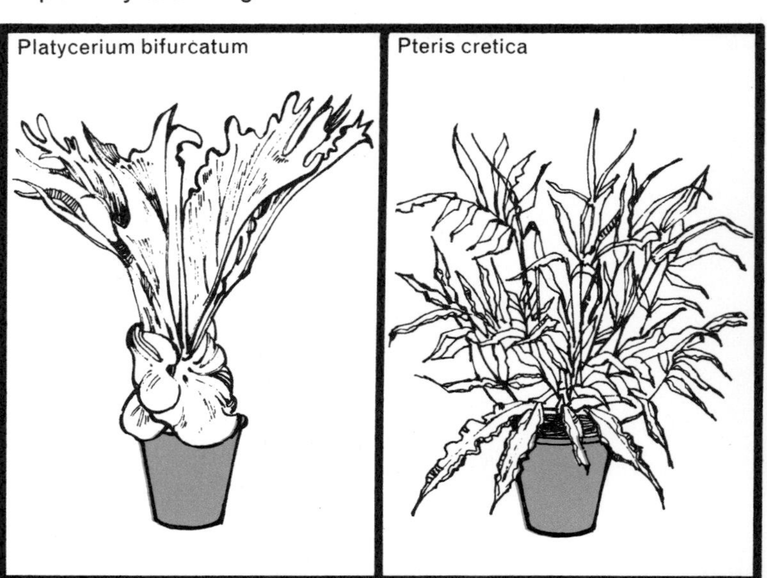

Platycerium bifurcatum | Pteris cretica

Filmies, Selaginellas

Though selaginellas are not ferns, they are closely related and thrive under similar conditions. They are ideal plants for the front of a greenhouse staging and will also grow happily amongst ferns

Filmy ferns have the most finely divided fronds of all, but require special conditions of quite dense shade and a very humid atmosphere. They are ideal for growing in Wardian cases, bottle gardens or plant cabinets in which the air is constantly saturated with moisture and the fronds are always damp.

There are numerous kinds, among the best being the Prince of Wales's Feather Fern, *Leptopteris superba*; the Killarney Fern, *Trichomanes radicans* and the Tunbridge Wells Fern, *Hymenophyllum tunbridgense*.

Selaginellas also grow well in similar damp, quite cool, shady conditions. They are not ferns, but they are allied to them; they are flowerless, produce spores and have minute leaves which give them a moss-like appearance. Most of them are quite small plants which can be grown among the ferns or along the front of the greenhouse staging.

There are numerous kinds such as *Selaginella caulescens*, one of the taller ones which often exceeds 1 ft.; *S. cuspidata*, low and tufted, the leaves margined with white; *S. emmeliana*, taller, green but with a white variegated variety; *S. kraussiana*, which soon creeps out of its pots to cover staging or floor and *S. uncinata*, of trailing habit and blue-green in colour.

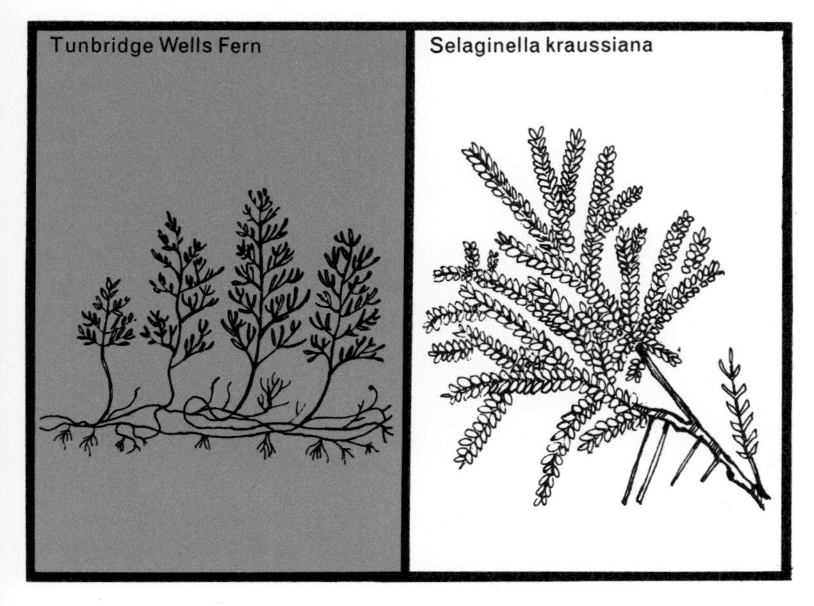

Keeping Ferns Healthy

A fernery, which can be established on a smaller scale than the one shown here, is the ideal way of maintaining the humid atmosphere which ferns enjoy and which discourages attacks from some pests

Ferns hardly ever suffer from disease, though occasionally mould appears on those with very finely divided fronds, and rusts can develop. There is little that can be done about rust except to cut off and burn any affected fronds, but moulds can be checked by letting the air dry a little and dusting the fern with flowers of sulphur. Care must be taken not to mistake for rust disease the natural spore clusters which form on fertile fronds and which are a means of propagation.

Most troublesome pest is thrips, a tiny, fast moving insect which sucks the sap in the fronds causing pale discolouration (especially along the veins) and curling of the fronds. This insect thrives in hot, dry air, so one way of countering it is to maintain a proper degree of humidity. Another is to fumigate occasionally with BHC (lindane).

Another fairly common pest is greenfly, which sucks sap and causes fronds to curl and become distorted. Use a BHC fumigant.

Mealy bugs are protected by a coating of a white waxy substance. Fortunately this makes them conspicuous and they can be picked off and destroyed as seen. Though malathion and diazinon sprays will kill mealy bugs they also damage ferns, so cannot safely be used.

Another mealy bug may attack the roots. If seen when re-potting it should be carefully hand picked, the roots washed in soapy water and then replaced in fresh soil.

Greenhouse Calendar

January

Bulbs of the impressive hippeastrums should be started into growth in this month

Take cuttings of late-flowering chrysanthemums, and particularly those of large-flowered types required for exhibition. Root the cuttings in sandy soil in a greenhouse or a propagator in a temperature of 13 to 18°C. (55 to 65°F.).

Cuttings of perpetual-flowering carnations can also be taken and rooted in similar conditions. Prepare root cuttings of Oriental Poppies, anchusas, herbaceous phlox, verbascums, Californian Tree Poppies (or romneyas), ramondas, haberleas and morisias.

In a greenhouse with a temperature of 15 to 18°C. (60 to 65°F.), sow seed of antirrhinums, begonias, cannas, gloxinias, streptocarpus, Scarlet Salvias and verbenas. Sweet peas can be sown in a slightly lower temperature.

Place tubers of begonias and gloxinias and achimenes rhizomes in a temperature of 15 to 18°C. (60 to 65°F.) to start them into growth. Bring hippeastrum bulbs into a similar temperature to start growth.

Transfer pots and bowls of daffodils, hyacinths and tulips, as well as *Iris* Wedgwood, from their frame or plunge bed into a moderately heated greenhouse to bring them into early flower.

Bring early flowering rock plants and small bulbs in pots and pans into an alpine house or any unheated, well ventilated and well lighted structure to protect the opening flowers from wind, rain and frost.

February

Cacti and succulents are always fascinating. Many can be raised from seed sown this month

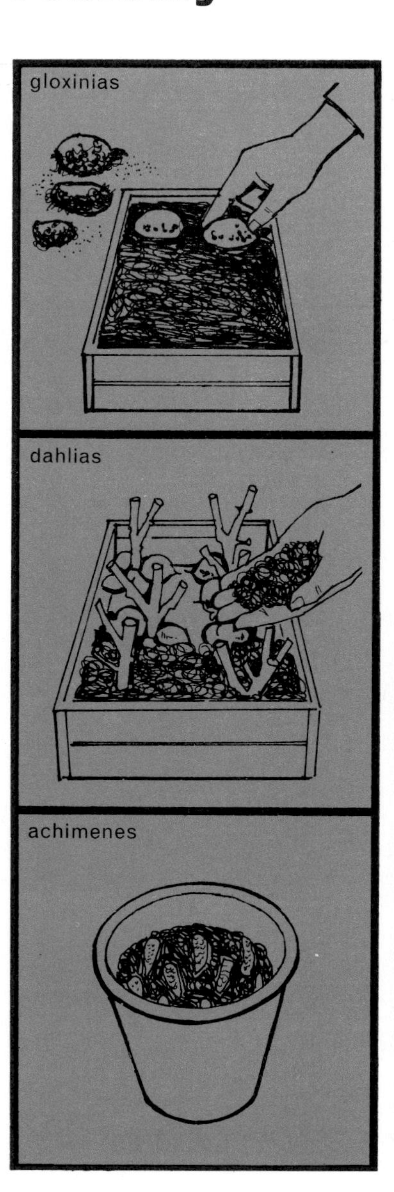

Continue all the work recommended for January.

Start more dormant tubers of begonias, gloxinias and achimenes.

Place dormant dahlia tubers in large pots or boxes or pack them close together in a greenhouse border surrounded and just covered with soil. Maintain a temperature of 13 to 15° C. (55 to 60° F.) to start the tubers into growth so that cuttings may be taken when the new shoots are 2 to 3 in. in length.

Continue to take cuttings of chrysanthemums, particularly those required to provide cut flowers in autumn and winter or to make large decorative plants at that time of the year.

Prune *Luculia gratissima* and *Lippia citriodora*, the former by shortening the previous year's growth to a few inches, the latter by cutting back almost to the base. Most other greenhouse shrubs (not in flower) should only be pruned enough to keep them tidy.

Gradually reduce the water supply to poinsettias so that the plants can have a short rest before being pruned and restarted to provide cuttings for a fresh stock.

Sow freesias in a temperature of 15 to 18° C. (60 to 65° F.). If sown thinly in pots the seedlings can grow on undisturbed to flower in autumn and winter ahead of plants grown from summer-potted corms.

Sow seeds of cacti in a similar temperature and do not prick out early but wait until the little plants have formed good root systems.

The end of the month is a good time to sow seed of the bushy Charm chrysanthemums

In a greenhouse or propagating frame with a temperature of 15 to 18°C. (60 to 65°F.), sow seeds of many half-hardy annuals and bedding plants. These can include antirrhinums, begonias, cannas, impatiens, petunias, Scarlet Salvias, salpiglossis and verbenas. Towards the end of the month, sow dahlias and charm and cascade chrysanthemums.

Also sow various summer-flowering greenhouse plants, including celosias and Cockscombs, *Clerodendrum fallax,* greenhouse begonias, gloxinias and streptocarpus.

Prick out any seedlings from January sowings that are sufficiently large to be handled. Often seedlings transplant most successfully when they have their first seed leaves and before they begin to grow the later leaves which are usually different in character. But with some very small seedlings, such as those of begonias and antirrhinums, this is just not practicable and it is better to wait until the seedlings have their first characteristic leaves.

Prune greenhouse plants that are to be trained as climbers in the house or are simply to be grown as well-shaped specimens. In the first category are bougainvilleas, the blue plumbago (Cape Leadwort) and ivy-leaved pelargoniums; and in the second, bouvardias, gardenias, fuchsias and zonal pelargoniums.

March

Any plants, such as these codiaeums, which are pot-bound should now be re-potted

Re-pot greenhouse plants that have become pot-bound, are in need of a change of soil or need starting into growth. These will include ornamental varieties of asparagus, codiaeum (crotons), coleus, cacti and other succulents, ferns, fuchsias, palms, pelargoniums (geraniums), smilax and most house plants that are not actually in flower at the time.

Pot rooted cuttings of chrysanthemums and perpetual-flowering carnations and move earlier plants, that have already filled their first small pots with roots, into larger pots.

Continue to take cuttings of chrysanthemums and perpetual-flowering carnations. Early March is a good time to take cuttings of early-flowering chrysanthemums to be planted out of doors in May.

Take cuttings of dahlias as shoots reach a suitable length (2 to 3 in.). Pot earlier dahlia cuttings singly in 3-in. pots as soon as they are well rooted, and grow on in a light greenhouse in a temperature of 12 to 15°C. (55 to 60°F.).

Also take cuttings of fuchsias, heliotropes and pelargoniums and root in a propagating frame or box in the greenhouse. Pot on cuttings of any of these taken the previous summer or early autumn. Pinch out the growing tips as necessary to induce bushy plants.

Re-pot any orchids that are starting into growth. Large plants can be divided. Old pseudobulbs (back bulbs) are best discarded.

Greenhouse Calendar

March

Abutilon Fireball flowers in the spring. Take cuttings this month and root at 15° C. (60° F.)

Take cuttings of winter-flowering begonias, such as *Begonia* Gloire de Lorraine, also of poinsettias, and root in a propagating frame with bottom heat and a temperature of 18°C. (65°F.).

Also take cuttings of various greenhouse plants that are most conveniently reproduced in this way. These may include abutilons, bouvardias, coleus, daturas, gardenias, jacobinias, lantanas, and the double-flowered tropaeolums. All will root readily in a propagating frame and a temperature of 15° C. (60° F.).

Stop young perpetual-flowering carnation plants by breaking out the topmost joint of each plant when it has made about seven pairs of leaves. This will encourage side-shoots to develop and lead to the bushy plants which are so desired.

Divide large roots of canna, potting the divisions in the smallest pots that will contain them comfortably, and grow them on in a temperature of 15°C. (60°F.).

Pot smithianthas, one stolon in each 4- or 5-in. pot, and start into growth in a temperature of 18°C. (65°F.).

Pot September-sown schizanthus into 5- or 6-in. pots in John Innes No. 2 Potting Compost or a peat compost. Grow in a light position at a temperature of 10 to 15°C. (50 to 60°F.).

Re-pot anthuriums, also any house plants and ferns that need it.

Sow seed of browallia and exacum for late summer flowers.

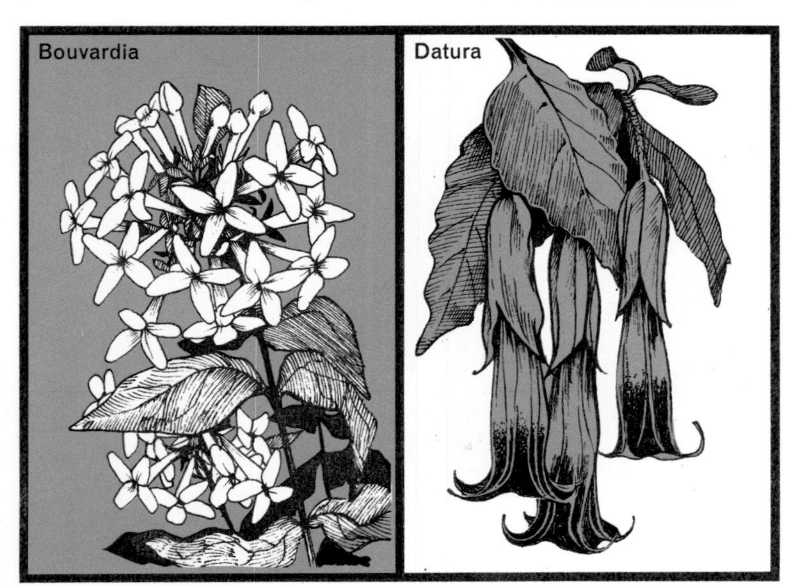

Bouvardia

Datura

April

This is a good time for a major re-potting of all the plants which need it

Sow winter-flowering primulas of all kinds including *Primula malacoides, P. obconica* and *P. sinensis*. Germinate in a temperature of about 15°C. (60°F.).

Sow annuals, both hardy and half-hardy, for growing on as flowering pot plants in the greenhouse. Particularly suitable kinds are *Begonia semperflorens*, annual carnations, exacum, gazanias, nemesias, petunias, salpiglossis, *Limonium suworowii* (statice), ten-week stocks, verbenas and zinnias.

Continue to take cuttings of chrysanthemums and dahlias. Pot rooted cuttings singly and remove earlier potted cuttings of these plants to a frame to be hardened off. Chrysanthemums are hardier than dahlias and may therefore go into a frame a week or so earlier.

Pot on rooted cuttings of perpetual-flowering carnations, chrysanthemums, fuchsias, pelargoniums (geraniums) and other greenhouse plants into larger pots as they fill their small pots with roots.

Re-pot any old plants that appear overcrowded or pot-bound, i.e. with roots tightly wound round one another in the pot.

Re-pot cacti that have outgrown their pots, but annual re-potting is unlikely to be necessary.

Take cuttings of fuchsias and pelargoniums (geraniums) if stock is short or late flowering plants are required.

Primula malacoides

Primula obconica

May

Calceolarias may need staking and tying to support the weight of flowers

Perpetual-flowering carnations are planted into their final pots this month

Pot on begonias and gloxinias as they fill their pots with roots, and shade from strong sunshine.

Continue to take cuttings of winter-flowering begonias and poinsettias. Pot up singly any earlier cuttings that are well rooted and grow on in a temperature of 15 to 18°C. (60 to 65°F.).

Re-pot evergreen azaleas and camellias when they have finished flowering. A peat-based compost suits these plants well.

Gradually reduce the water supply to freesias and lachenalias that have been flowering during the winter or early spring so that they can rest for a while, but keep young freesias growing by watering them freely.

Pot on young cyclamen plants as they fill the smaller pots with roots.

Start late batches of achimenes and hippeastrums to give a succession of flowers.

Stake and tie greenhouse calceolarias carefully as they come into flower. The dwarf or nanus varieties do not require staking.

Fumigate the greenhouse with a suitable insecticide to keep down greenflies and other pests.

Continue to stop perpetual-flowering carnation cuttings.

Sow schizanthus to flower in summer and early autumn.

Start to harden off all plants that are to go outdoors in summer. Many plants, including greenhouse azaleas, can now be kept in frames, except in cold areas.

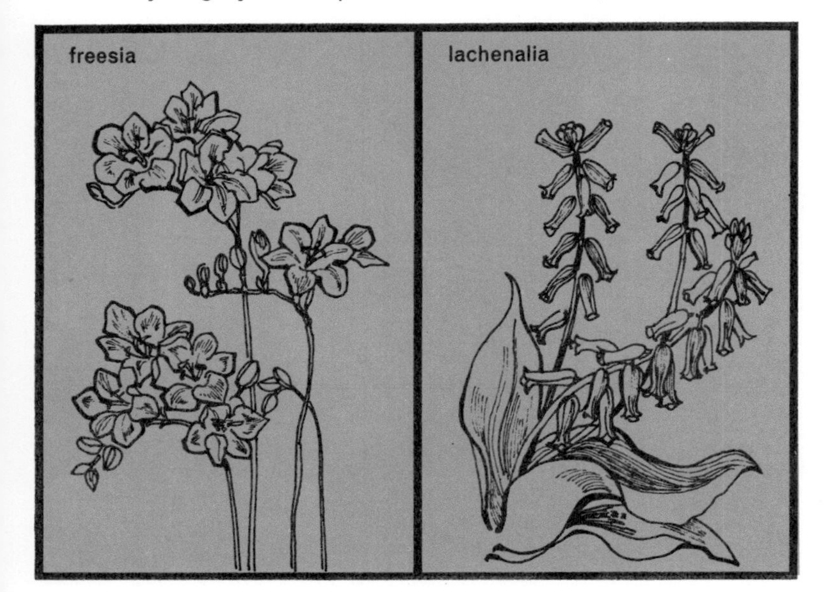

Shade greenhouses from strong sunshine unless plants are being grown that need as much sun heat as possible e.g. cacti and other succulents.

Give free ventilation by day but close ventilators early to trap sun heat, especially if there is a risk of frost.

Water pot plants freely and start to feed established plants with weak liquid manure.

Fumigate occasionally with nicotine smoke generators to kill greenflies, whiteflies etc.

Pot on spring-rooted cuttings of pelargoniums, fuchsias, winter-flowering begonias, poinsettias etc. as they fill the smaller pots with roots.

Pot late-flowering chrysanthemums into the pots in which they will grow on all the summer and flower in the autumn or winter. Place a bamboo cane in each pot and then stand the plants outdoors in a sheltered but sunny place on a firm ash or gravel base. Keep well watered.

Pinch out the tips of chrysanthemum plants grown for autumn or winter cut flowers or as decorative plants.

Insert cuttings of dwarf chrysanthemums to make small pot plants.

Also pot perpetual-flowering carnations into their final pots, usually 6 or 7 in. in diameter. Break out ends of side growths when each has made about 8 pairs of leaves.

Greenhouse Calendar

May

The summer-flowering tuberous begonia Roy Hartley has large flowers of excellent quality

Gradually reduce the amount of water given to Arum Lilies, lachenalias, nerines and freesias that have flowered in winter and allow the plants to rest for a time. Place nerines in the sunniest part of the greenhouse so that they get a good baking.

Sow cinerarias in a greenhouse or frame. No artificial heat is necessary at this time of the year.

Prick out seedlings of winter-flowering primulas, such as *Primula malacoides, P. obconica* and *P. sinensis*. Grow on in a shaded greenhouse or frame and water fairly freely. Make further sowings for successive flowers.

Plant up summer-flowering begonias and gloxinias in the pots in which they will flower. Grow on in a shaded greenhouse with only sufficient artificial heat to prevent the temperature falling below 10°C. (50°F.) at night.

Fill hanging baskets with geraniums, trailing lobelias and campanulas, pendulous begonias etc. and suspend them in the greenhouse for a few weeks to get established before hanging them outdoors.

Complete the hardening off of all plants that are to go out of doors in summer. Frame lights can be removed on all fine days.

Prune plants of *Euphorbia splendens* that are getting too tall and are not well branched.

Sow browallia to provide plants to flower in autumn and winter.

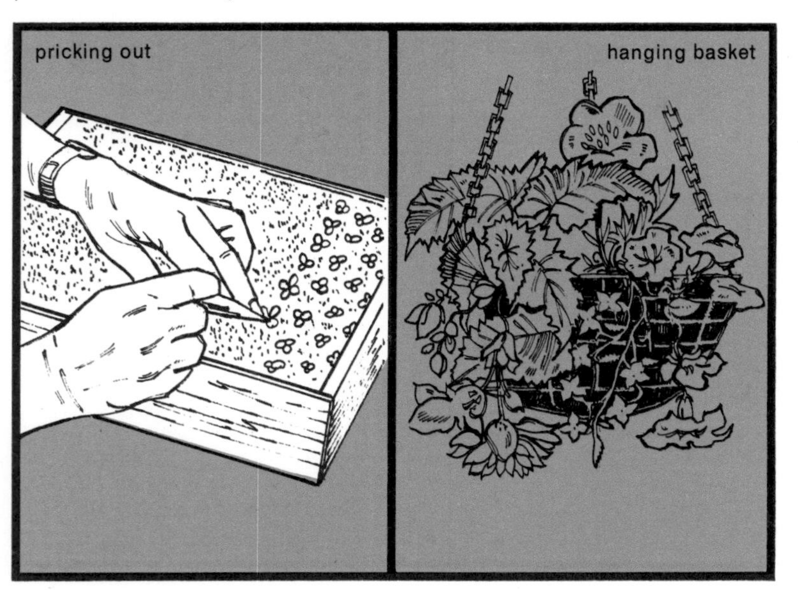

June

Cinerarias flower in winter and spring but the seed should be sown this month

Sow calceolarias and make further sowings of cinerarias and of winter-flowering primulas. This is a very good time to sow seed of *Primula malacoides*.

Prick out seedlings of cinerarias and winter-flowering primulas from earlier sowings, and pot the most forward seedlings singly in 3-in. pots.

Complete the potting of late-flowering chrysanthemums and perpetual-flowering carnations.

Stand out any remaining chrysanthemums on a gravel or ash base in a sheltered place out of doors. Perpetual-flowering carnations may be placed in a frame for the summer if the greenhouse is required for other purposes.

Many plants only require greenhouse protection when there is danger of frost, and can stand out of doors all summer.

See that greenhouses are well shaded on bright, sunny days, give ample ventilation to keep the temperature from rising too high, and damp down paths and stagings to maintain plenty of moisture in the atmosphere.

Pot on begonias and gloxinias as the smaller pots become comfortably filled with roots.

Stop all chrysanthemum plants that have not already been so treated by pinching out the top of each plant to make it branch. Some late-flowering varieties can be stopped twice; for such information consult one of the specialist chrysanthemum catalogues.

Greenhouse Calendar

July

This regal pelargonium is Summertime. It is one of many varieties for greenhouse culture

Pot-grown chrysanthemums must now be kept well watered. Feed every 10 to 14 days

Move young cyclamen plants into 5-in. pots as soon as they have filled the smaller pots with roots. Keep the plants growing all summer in a cool frame or a shady greenhouse and do not let them get dry at any time. Remove old cyclamen plants that have already flowered to a shady frame and give very little water until August.

Re-pot pot-grown auriculas. When doing this, examine them for root aphids (small grey lice on the roots), and if any are seen, wash the roots in a solution of an insecticide such as lindane, derris or malathion.

Many woody-stemmed plants, both outdoor and indoor, can be increased by cuttings taken between June and August and rooted in a propagating frame or box or in polythene bags in a greenhouse. Prepare cuttings from firm young shoots and insert a few each month as some may root more freely than others.

Towards the end of the month, when regal pelargoniums are no longer flowering freely, cut the plants back and place them in a frame or sunny place out of doors for a few weeks.

Fumigate occasionally with nicotine or other suitable smoke generators to keep down pests. Use azobenzene if red spider mite is troublesome.

While plants of *Solanum capsicum* are in flower syringe daily with water to encourage fruits to set.

re-potting cyclamen

spraying chrysanthemums

damping down

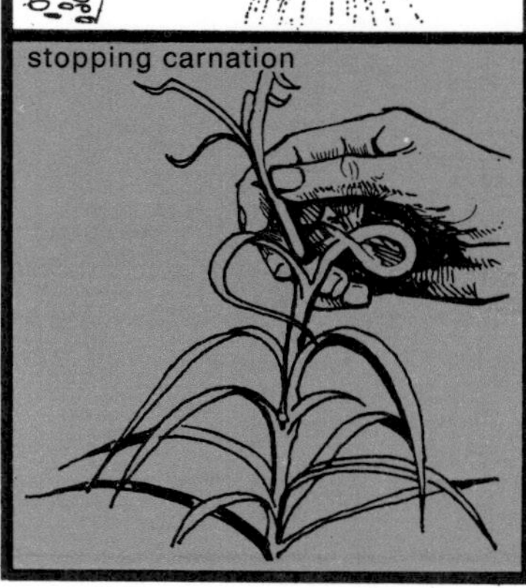

stopping carnation

Water pot-grown chrysanthemums well and feed every 10 to 14 days. Spray with a good insecticide to keep them free of aphids and other pests.

Ventilate greenhouses freely in sunny weather. Many plants will require shading from direct sunlight and, except in the case of most succulents (including cacti) and carnations which prefer a dry atmosphere, water paths and syringe the walls and staging with water to maintain plenty of moisture in the air.

Stop young perpetual-flowering carnation plants for the last time by breaking out the end of each shoot.

Cuttings of many plants, including hardy trees and shrubs as well as those commonly grown in greenhouses, can be rooted in a close frame or under mist. Young growths that are becoming firm at the base make the best cuttings. Most will root more rapidly and certainly with bottom heat, i.e. the soil warmed from below so that it is at least as warm, possibly a little warmer, than the mean temperature of the air. The cut should be made through a node (where the leaf joins the stem), and the cut surface dipped into hormone rooting compound.

Fumigate the greenhouse occasionally to keep down pests. If whitefly appears several fumigations may be necessary to cope with the further batches of adult 'flies' that will emerge from the scales even after fumigation.

Greenhouse Calendar

July

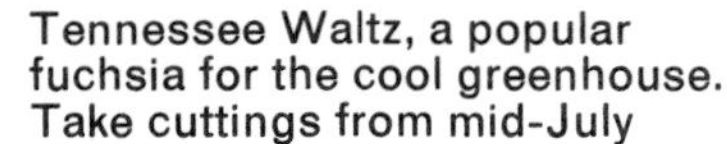
Tennessee Waltz, a popular fuchsia for the cool greenhouse. Take cuttings from mid-July

Prick off winter-flowering primulas, cinerarias and calceolarias and pot singly those that have already been pricked off and have grown into sturdy little plants. Earlier batches already in pots should be moved on to larger sizes as soon as they fill the smaller pots with roots.

Make further sowings of calceolarias, cinerarias and *Primula malacoides* for successional flowering. At this season germination is just as good in a frame as in a greenhouse.

Pot on all young plants grown from cuttings, divisions etc. as they fill their smaller pots and from the middle of the month take cuttings of greenhouse fuchsias.

Keep nerines in the sunniest part of the greenhouse and do not water them. Keep Arum Lilies rather dry for a few weeks before re-potting them in August. Stand regal pelargoniums outdoors in a sunny, sheltered place for a few weeks and water very sparingly.

Shade begonias and gloxinias well, water freely and feed them every 10 days or so with weak liquid manure. Stake and tie the flowering stems of begonias carefully to support the heavy blooms.

Take care not to allow the air to get too dry for those plants that thrive best in a moist atmosphere. This includes almost all foliage plants, ferns and orchids, as well as begonias and gloxinias. Keep paths and staging gravel wet, syringe between pots or install a humidifier.

August

This is a good time to begin planting up bowls of narcissus bulbs for winter flowering

Some flower buds may appear on late-flowering chrysanthemums during August and must be reduced to one per stem if large flowers are required.

Continue to ventilate greenhouses freely, to syringe frequently and to shade those plants that do not like intense heat or strong light. In small greenhouses it may be necessary at times to leave the doors open to keep the temperature down.

As the earliest batches of achimenes, tuberous-rooted begonias and gloxinias come to the end of their flowering period, gradually reduce the water supply and allow them to die down naturally.

Pot on winter-flowering primulas, cinerarias and calceolarias when they fill the smaller pots with roots, keeping them in a frame if possible as these plants do not like high temperatures.

Pot hyacinths, narcissi and other bulbs required for winter flowering. Bulbs for indoor flowering should also be planted now in bowls of bulb fibre. Keep them all in a cool, dark place for the time being or plunge pot-grown bulbs (but not bowl-grown ones) under 4 in. of sand or peat out of doors.

Pot veltheimias and keep in a sunny frame.

Sow winter-flowering stocks, also exacum and schizanthus to flower early next year.

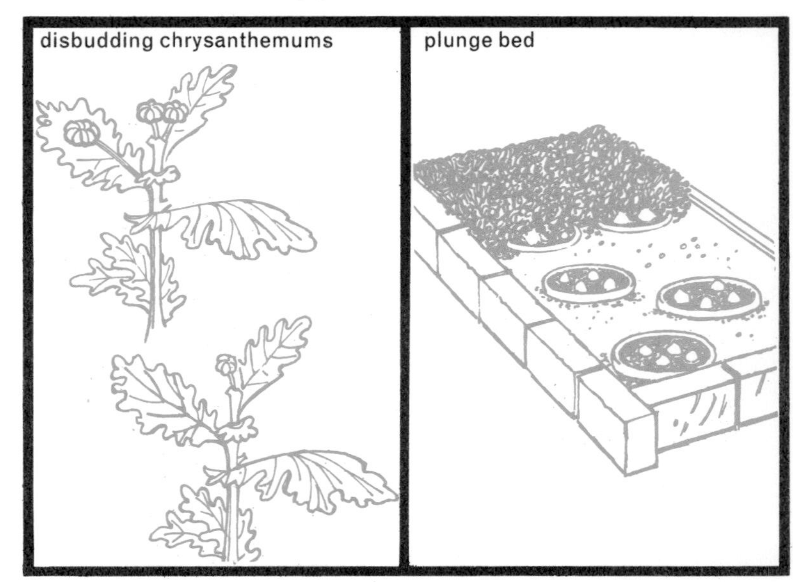

Greenhouse Calendar

Lachenalia bulbs should be potted during August to flower in winter and early spring

Cyclamen are among the most decorative of plants. Pot seedlings singly this month

September

Pot Arum Lilies or re-start those that have been resting during the summer by gradually increasing the quantity of water given to them.

Re-start nerines or purchase and pot bulbs in John Innes No. 1 Potting Compost, keeping the bulbs well up in the pots so that much of the bulb is exposed.

Re-pot freesia corms early in the month but keep them in a frame or a really cool greenhouse for the time being. Pot lachenalias or plant them in hanging baskets.

Sow cyclamen for flowering in the greenhouse the following year. Cyclamen seeds are often slow and irregular in germinating so prick out the seedlings as soon as they become sufficiently large, but do not discard the seed pans for several months. Re-pot in a shady frame old cyclamen corms that have been resting during the summer.

Cut back regal pelargoniums, start to water the plants more freely and when new shoots are 3 or 4 in. long take some cuttings.

Continue to fumigate the greenhouse occasionally. This is a month when whitefly, red spider mites and mealy bug can be troublesome. Azobenzene aerosols give good control of red spider mites, and mealy bugs may be sponged off with derris.

Take cuttings of zonal and ivy-leaved pelargoniums, also of fuchsias, gazanias, lampranthus (mesembryanthemums) and other plants of doubtful hardiness.

If greenhouses can be completely emptied they should be fumigated by burning sulphur in them, but the fumes are fatal to plants which must not be returned to the greenhouse until all traces of smoke have dispersed.

Make certain that the heating apparatus is functioning properly and carry out any necessary repairs before the winter. If the apparatus is thermostatically controlled check the operation and accuracy of the thermostat.

Cinerarias are liable to be attacked by leaf miner maggots which leave snaky white tunnels in the leaves. If these are seen, feel for the maggots in the leaves and kill them with the point of a penknife or spray them with an insecticide such as BHC or trichlorphon.

Sow schizanthus for spring and early summer flowering, if not already done in August. Also sow any annuals required for use as pot plants in the greenhouse.

Pot singly any seedlings of cyclamen that appear from the August sowings but do not discard the seed pans or disturb the surface more than is essential as germination is often slow and irregular.

Continue to take cuttings of zonal and ivy-leaved pelargoniums, gazanias, lampranthus (mesembryanthemums), fuchsias.

Cuttings of coleus root readily in September and young plants are often easier to over-winter than old plants.

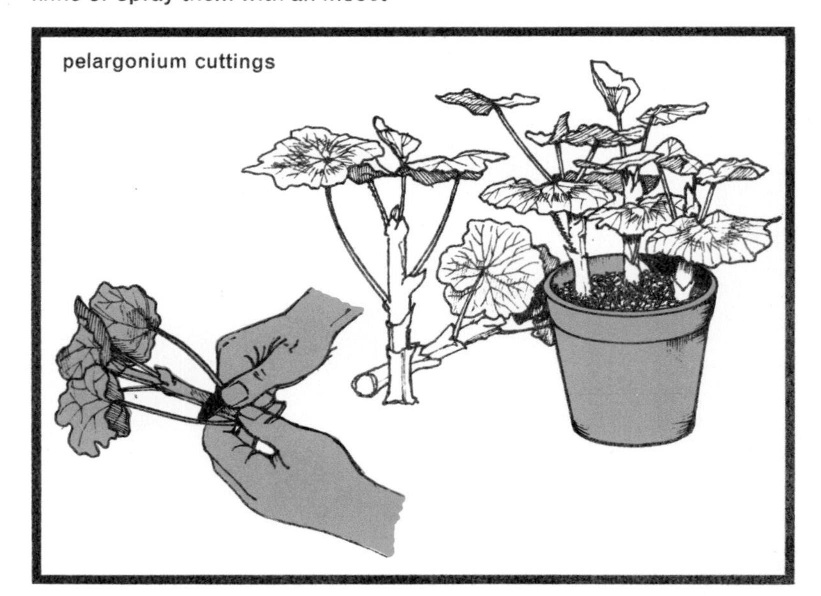

Greenhouse Calendar

September

The most popular lily for greenhouse cultivation is *Lilium longiflorum*. Pot the bulbs now

Many plants need all the light they can get in September and little shading is likely to be required. Syringing and damping down are also probably unnecessary except for a few moisture-loving plants.

Continue to pot bulbs for the greenhouse. This is a good time to pot *Lilium longiflorum* and other greenhouse lilies if the bulbs can be acquired so early. Pot a further batch of freesia corms for successive flowering.

Gradually reduce the amount of water given to begonias, gloxinias and hippeastrums so that they may ripen their growth. Old cyclamen corms re-potted in August will need increasing amounts of water as they start vigorously into growth. Nerines will also require more water as their flower stems grow.

Pot on winter-flowering primulas, calceolarias and cinerarias into the pots in which they will flower. Towards the end of the month bring them in from the frame where they have passed the summer back to a light, frost-proof greenhouse.

Also bring into the greenhouse any other tender plants that have been out of doors or in frames, including perpetual carnations, late-flowering chrysanthemums, Winter Cherries (solanums) and Indian azaleas. House plants which are returned indoors should have large leaves sponged of any dust.

October

Primula obconica is available in a range of colours. It flowers in winter and for most of the year

If not already done in September, bring in late-flowering chrysanthemums and any other greenhouse plants that have been standing out of doors during the summer.

Gradually reduce the water supply to tuberous-rooted begonias, gloxinias, achimenes and ordinary hippeastrums so that by the end of the month they can be dried off completely for the winter.

Finish any potting on of young plants that was not completed in September, including winter-flowering primulas, cinerarias, calceolarias and cyclamen, but move the plants only if their pots are uncomfortably full of roots.

See that plants have as much light and as much room as possible and be careful when watering not to wet the leaves and crowns unnecessarily. Fertilisers should not be used at all during the autumn and winter months.

Continue to pot tulips, hyacinths and other bulbs for the greenhouse, including hippeastrums specially prepared for winter flowering. Bring in the earliest potted batches if they have formed plenty of roots and keep in full light in a temperature of 10 to 15° C. (50 to 60° F.).

Pot annuals sown in August and September, also winter-flowering stocks.

Disbud perpetual-flowering carnations, retaining only one flower bud per stem.

Pot a further batch of freesia corms to provide a succession of flowers.

Greenhouse Calendar

November

Varieties of *Camellia japonica* make good pot plants for the cool greenhouse

Use sufficient artificial heat to maintain the necessary minimum temperatures for the plants being grown. Ventilate only by day when the weather is bright and there is sufficient sunshine to maintain a temperature of 13 to 15° C. (55 to 60° F.) and water carefully, only when it is really necessary.

Bring successive batches of bulbs in pots from frames or outdoor plunge beds into the greenhouse, but only when they have filled their pots fairly well with roots. At first the night temperatures should not be above 13° C. (55° F.) – less will do.

Pot astilbes, dicentras and other herbaceous plants to be gently forced in spring, also roses and shrubs such as Indian azaleas, camellias, hydrangeas, lilacs, genistas and deutzias.

Keep established Indian azaleas well watered and raise the temperature a little to around 15° C. (60° F.) if early flowers are wanted.

Continue to remove all diseased leaves and stems and to dust such plants with flowers of sulphur.

To support the lengthening flower stems of freesias, place a few sticks in the pots and loop raffia or fillis (soft string) around them.

Cut down indoor chrysanthemums as they finish flowering, keeping only enough plants of each variety for propagation purposes.

Reduce the water supply to smithianthas until quite dry and then store them like achimenes.

December

Poinsettias are a colourful sight this month. Red, pink or white varieties are available

Begin taking cuttings of perpetual-flowering carnations, choosing short sideshoots from about half way up the flowering stems.

Continue to cut back chrysanthemums as they finish flowering and to bring in successive batches of bulbs in pots to be forced gently in the greenhouse. Keep a close watch on temperatures, particularly now that nights are very cold. Many greenhouse plants suffer if temperatures fall below 7° C. (45° F.). If necessary, use auxiliary heating, such as an oil stove, if the main apparatus is inadequate.

Keep the air as dry as possible and water rather sparingly as plants suffer more from cold when overwet. Remove any diseased leaves and dust infected plants with flowers of sulphur or with thiram.

Give flowering plants such as winter begonias, cinerarias, lachenalias and primulas the lightest positions in the house and keep the glass as clean as possible at this time of the year.

Do not let poinsettias and saintpaulias get chilled while in flower.

Prune oleander by shortening to a few inches all growth made in the past year. Prune greenhouse roses, including any permanently planted climbing roses such as Maréchal Niel and Niphetos.

Examine tubers and bulbs in store including achimenes, begonias, cannas, gloxinias, hippeastrums and smithianthas and dust with flowers of sulphur if there is any sign of decay.

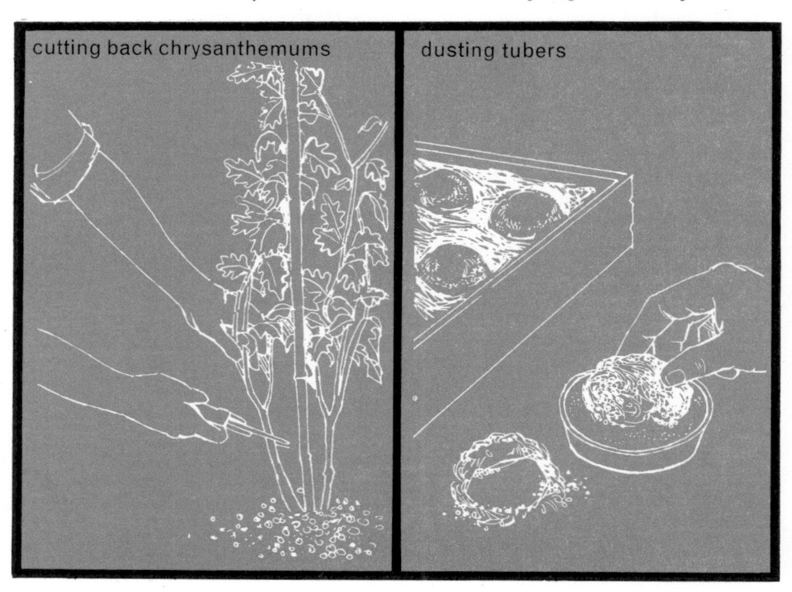

Index

Abbreviation: p = colour illustration or line drawing